Rain, Snow, and Ice Loads

Time-Saving Methods Using the 2018 IBC and ASCE/SEI 7-16

David A. Fanella,
Ph.D., S.E., P.E., F.ACI, F.ASCE, F.SEI

New York Chicago San Francisco
Athens London Madrid
Mexico City Milan New Delhi
Singapore Sydney Toronto

Library of Congress Cataloging-in-Publication Data

Names: Fanella, David Anthony, author.
Title: Rain, snow, and ice loads : time-saving methods using the 2018 IBC and ASCE/SEI 7-16 / David A. Fanella.
Description: New York : McGraw Hill, [2021] | Includes bibliographical references and index. | Summary: "A concise, visual guide for engineers designing structures to withstand rain, snow, and ice loads, this book will present explanations and workflows for the 20% of the building code that engineers use 80% of the time"—Provided by publisher.
Identifiers: LCCN 2020017597 | ISBN 9781260461527 (paperback ; acid-free paper) | ISBN 9781260461534 (ebook)
Subjects: LCSH: Snow loads. | Hydrostatic pressure. | Rainstorms. | Ice. | Building—Cold weather conditions—Standards. | Building, Stormproof—Standards.
Classification: LCC TA654.4 .F36 2021 | DDC 624.1/76—dc23
LC record available at https://lccn.loc.gov/2020017597

**Rain, Snow, and Ice Loads:
Time-Saving Methods Using the 2018 IBC and ASCE/SEI 7-16**

1 2 3 4 5 6 7 8 9   CD   25 24 23 22 21 20

ISBN 978-1-260-46152-7
MHID 1-260-46152-1

The pages within this book were printed on acid-free paper.

Sponsoring Editor
Ania Levinson

Editorial Supervisor
Donna M. Martone

Acquisitions Coordinator
Elizabeth Houde

Project Manager
Parag Mittal,
Cenveo® Publisher Services

Copy Editor
Cenveo Publisher Services

Proofreader
Cenveo Publisher Services

Production Supervisor
Pamela A. Pelton

Composition
Cenveo Publisher Services

Art Director, Cover
Jeff Weeks

Contents

About the Author

David. A. Fanella is Senior Director of Engineering at the Concrete Reinforcing Steel Institute where his main responsibility is creating educational material for structural engineers, including publications, design aids, and webinars. He has over 30 years of experience in a wide variety of low-, mid-, and high-rise buildings and other structures and has authored numerous books and technical papers through the years, including two editions of *Reinforced Concrete Structures, Analysis and Design*. David is a licensed Structural Engineer and Professional Engineer in Illinois, and is a Fellow of the American Concrete Institute, the American Society of Civil Engineers, and the Structural Engineers Institute. He is active in many professional organizations, including membership on ASCE/SEI 7 and ACI Committees. He is also past President and past Board Member of the Structural Engineers Association of Illinois.

Preface

This publication provides structural engineers, educators, students, and other design professionals a concise, visual guide to the determination of structural loads due to rain, snow, and ice. The intent is to present the provisions in the 2018 *International Building Code* and ASCE/SEI 7-16 *Minimum Design Loads and Associated Criteria for Buildings and Other Structures* in a manner that is easy to understand and apply. This is achieved by utilizing step-by-step methods including numerous figures, tables, flowcharts, and design aids.

Examples in both inch-pound and metric (S.I.) units illustrate the proper application of the code provisions and follow the step-by-step methods outlined throughout the publication. Section, figure, table, and equation numbers from the code and this publication are given in the right-hand margin of the examples for easy reference.

In short, rain, snow, and ice loads can be determined simpler and faster using the procedures in this publication.

For further online information on this topic, please go to https://www.mhprofessional .com/RainSnowIceLoads

David A. Fanella

CHAPTER 1
Introduction

1.1 Overview

The purpose of this publication is to assist in the proper determination of rain, snow, and ice loads in accordance with the 2018 edition of the *International Building Code©* (IBC©) [Ref. 1; see Chap. 5 of this publication for a list of references] and the 2016 edition of ASCE/SEI 7 *Minimum Design Loads and Associated Criteria for Buildings and Other Structures* (Ref. 2). The main goal is to streamline the load determination process by providing straightforward, step-by-step procedures enhanced by numerous design aids, figures, and flowcharts, which provide a roadmap through the numerous code requirements.

Design professionals will appreciate the simplicity and thoroughness of the content and will find the "how to" methods of load determination useful in everyday practice. Worked-out examples illustrate the proper application of the code requirements and follow the step-by-step procedures noted above; these examples are a valuable resource for individuals studying for licensing exams, undergraduate and graduate students, and others involved in structural engineering.

Readers interested in the background, history, and design philosophy of the code requirements for rain, snow, and ice loads can find detailed information and references in the commentary of Ref. 2.

1.2 Scope

Throughout this publication, section numbers from the IBC are referenced as illustrated by the following: Section 1611 of the IBC is denoted as IBC 1611. Similarly, Section 8.3 of ASCE/SEI 7-16 is referenced as ASCE/SEI 8.3.

Chapter 2 contains methods to calculate design rain loads in accordance with IBC 1611 and ASCE/SEI Chapter 8. Requirements for ponding instability and ponding loads are also covered. Examples demonstrate the calculation of design rain loads for roofs with edge overflow, standpipe drainage systems, overflow dam drainage systems, and scuppers.

Methods to calculate design snow loads in accordance with IBC 1608 and ASCE/SEI Chapter 7 are given in Chap. 3. Design snow load examples include buildings with the following: monoslope roofs with and without ice dam loads; gable roofs with various slopes; curved roofs; sawtooth roofs; upper and lower roofs, including sliding snow; an adjacent separated building; canopies; parapets; and rooftop units. Also included are examples for existing buildings with adjacent new buildings and elements in open-frame equipment structures.

Given in Chap. 4 are methods to determine atmospheric loads due to freezing rain in accordance with IBC 1614 and ASCE/SEI Chapter 10. Included are examples on ice weight and wind-and-ice load determination for a tank, a chimney, and a solid free-standing sign.

Chapter 5 contains the references cited in this publication.

Both inch-pound and S.I. units are used throughout this publication, including in the equations, figures, tables, flowcharts, and examples. In the examples, calculations are performed independently using both sets of units; in other words, the calculations are not performed in one set of units and then converted to the other. Thus, in some cases, the numerical results in inch-pound units do not "exactly" convert to the corresponding numerical results in S.I. units or vice versa.

CHAPTER 2

Rain Loads

2.1 Overview

This chapter contains methods to calculate design rain loads, R, in accordance with IBC 1611 and ASCE/SEI Chapter 8. The total rain load is equal to (1) the load created by the amount of accumulated rainwater on a roof assuming the primary drainage system for that portion is blocked plus (2) the uniform load caused by water that rises above the inlet of the secondary drainage systems at its design flow. Requirements for ponding instability and ponding loads are also covered.

2.2 Notation

A = tributary roof area plus one-half the wall area that diverts rainwater onto the roof (where applicable) serviced by a single drain outlet in the secondary drainage system, ft² (m²)

b = width of channel or closed scupper, in. (mm)

d_h = additional depth of water on the undeflected roof above the inlet of the secondary drainage system at its design flow (that is, the hydraulic head), in. (mm)

d_{h1} = known hydraulic head from ASCE/SEI Tables C8.3-1 and C8.3-2, in. (mm)

d_{h2} = hydraulic head to be determined by ASCE/SEI Equation (C8.3-3) for the specified secondary drain, in. (mm)

d_s = depth of water on the undeflected roof up to the inlet of the secondary drainage system when the primary drainage system is blocked (that is, the static head), in. (mm)

D = overflow dam or standpipe diameter, in. (mm)

= diameter of circular scupper, in. (mm)

D_1 = overflow dam or standpipe diameter for secondary (overflow) drain corresponding to d_{h1} for a given flow rate, Q, as shown in ASCE/SEI Tables C8.3-1 and C8.3-2, in. (mm)

D_2 = specified overflow dam or standpipe diameter for secondary (overflow) drain corresponding to d_{h2} for a given flow rate, Q, in. (mm)

h = height of channel or closed scupper, in. (mm)

i = design rainfall intensity, in./h (mm/h)

7

L_p = span of primary members, ft (m)

L_r = length of level roof edge that allows for free overflow drainage of rainwater when the roof edge is acting as the secondary drainage system, ft (m)

L_s = span of secondary members, ft (m)

Q = flow rate out of a single drainage system, gal./min (m³/s)

R = rain load on the undeflected roof, in lb/ft² (kN/m²). When the phrase "undeflected roof" is used, deflections from loads (including dead loads) are not considered when determining the amount of rain on the roof.

S = spacing of secondary members, ft (m)

β = roof rise in inches (mm) per unit run

2.3   Procedure to Determine Rain Load, *R*

A step-by-step procedure to determine the rain load, R, is given in Fig. 2.1. The sections of this publication referenced in Fig. 2.1 contain additional information needed to calculate R.

2.4   Rainfall Intensity, *i*

Design rainfall intensity, i, can be obtained by using the map in IBC Figure 1611.1 or by entering an address or the latitude and longitude of the site in Refs. 3 and 4 (see Table 2.1).

A 60-minute duration/100-year return period rainfall event is specified in IBC 1611 for the design of both the primary and secondary drainage systems. In ASCE/SEI 8.2, the design of the primary drainage system is to be based on a rainfall intensity equal to or greater than the 60-minute/100-year return period storm. Also, the design flow rate of the secondary (overflow) drains (including roof drains and downstream piping) or scuppers and their resulting hydraulic head, d_h, is to be based on a rainfall intensity equal to or greater than the 15-minute/100-year return period storm.

With everything else being equal, i for a 15-minute/100-year event is about 2.0 to 2.5 larger than i for a 60-minute/100-year return event, which means Q and R are larger for 15-minute/100-year events. It is always good practice to check with the local jurisdiction to ensure that the proper value of i is used in determining R.

2.5   Flow Rate, *Q*

The flow rate, Q, of rainwater through a single drainage system is determined by ASCE/SEI Equations (C8.3-1) and (C8.3-1.si):

$$Q = 0.0104 Ai \quad \text{(gal./min)} \tag{2.1}$$

$$Q = (0.278 \times 10^{-6}) Ai \quad \text{(m}^3\text{/s)} \tag{2.2}$$

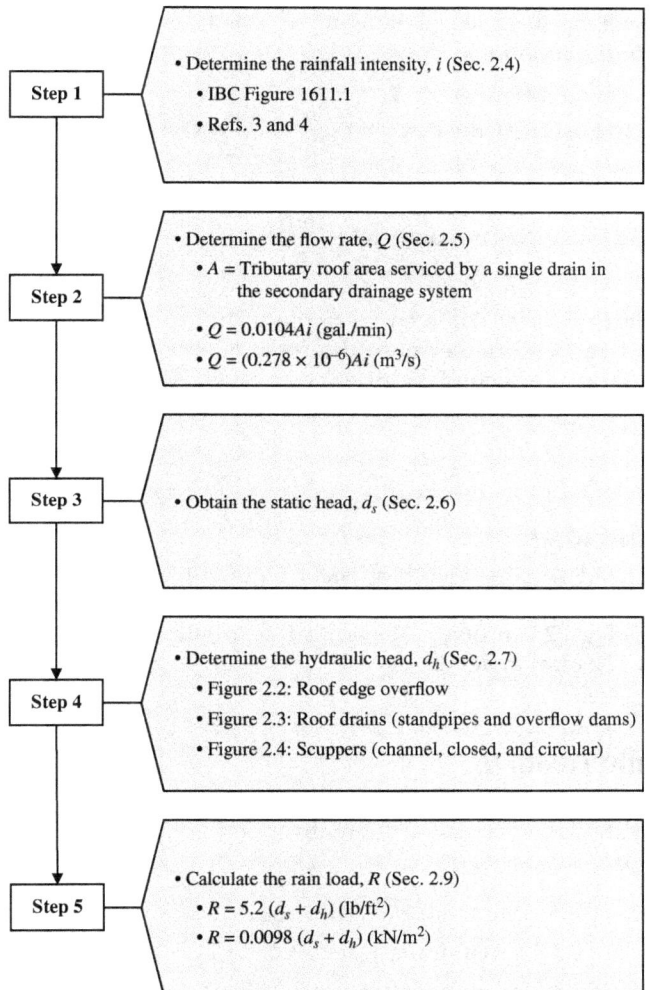

Step 1
- Determine the rainfall intensity, i (Sec. 2.4)
  - IBC Figure 1611.1
  - Refs. 3 and 4

Step 2
- Determine the flow rate, Q (Sec. 2.5)
  - A = Tributary roof area serviced by a single drain in the secondary drainage system
  - $Q = 0.0104Ai$ (gal./min)
  - $Q = (0.278 \times 10^{-6})Ai$ (m³/s)

Step 3
- Obtain the static head, d_s (Sec. 2.6)

Step 4
- Determine the hydraulic head, d_h (Sec. 2.7)
  - Figure 2.2: Roof edge overflow
  - Figure 2.3: Roof drains (standpipes and overflow dams)
  - Figure 2.4: Scuppers (channel, closed, and circular)

Step 5
- Calculate the rain load, R (Sec. 2.9)
  - $R = 5.2\,(d_s + d_h)$ (lb/ft²)
  - $R = 0.0098\,(d_s + d_h)$ (kN/m²)

Figure 2.1   Procedure to determine rain load, R.

Resource	Rainfall Event Parameters	
	Duration (minutes)	**Return Period (years)**
IBC Figure 1611.1	60	100
ASCE 7 Hazard Tool (Ref. 3)	15	100
	60	100
Precipitation Frequency Data Server (Ref. 4)*	15	100
	60	100

*Reference 4 also contains rainfall event parameters for events with durations and return periods other than those listed in this table.

Table 2.1   Resources to Obtain Design Rainfall Intensity, i

The constants in these equations are obtained based on the units associated with the variables in the equations:

$$\text{In Eq. (2.1): Constant} = \text{ft}^2 \times \frac{\text{in.}}{\text{h}} \times \frac{1 \text{ ft}}{12 \text{ in.}} \times \frac{7.48 \text{ gal.}}{\text{ft}^3} \times \frac{1 \text{ h}}{60 \text{ min}} = 0.0104$$

$$\text{In Eq. (2.2): Constant} = \text{m}^2 \times \frac{\text{mm}}{\text{h}} \times \frac{1 \text{ m}}{1{,}000 \text{ mm}} \times \frac{1 \text{ h}}{3{,}600 \text{ s}} = 0.278 \times 10^{-6}$$

The tributary area, A, is equal to the tributary roof area plus one-half the wall area that diverts rainwater onto the roof (where applicable) serviced by a single drain outlet in the secondary drainage system. Relatively large walls adjacent to roofs have the potential to divert substantial wind-driven rain flow down the wall to the roof.

2.6   Static Head, d_s

The static head, d_s, is the depth of water on the undeflected roof up to the inlet of the secondary drainage system and is determined in the design of the combined drainage system (see Figs. 2.2 through 2.4 below). It is usually specified to be in the range of 2 to 4 in. (51 to 102 mm) in depth.

2.7   Hydraulic Head, d_h

The hydraulic head, d_h, is related to Q and the type and size of the secondary drainage system. Methods to determine d_h are given in the following figures for three types of secondary drainage systems:

- Figure 2.2 for roof edge overflow (see ASCE/SEI C8.3)
- Figure 2.3 for roof drains (see ASCE/SEI C8.3)
- Figure 2.4 for scuppers (see ASCE/SEI C8.3 and Ref. 5)

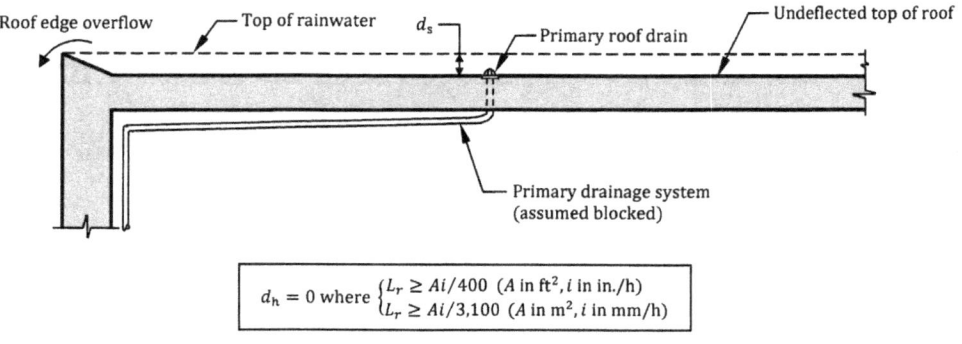

$$d_h = 0 \text{ where } \begin{cases} L_r \geq Ai/400 & (A \text{ in ft}^2, i \text{ in in./h}) \\ L_r \geq Ai/3{,}100 & (A \text{ in m}^2, i \text{ in mm/h}) \end{cases}$$

Figure 2.2   Determination of d_h—Roof edge overflow.

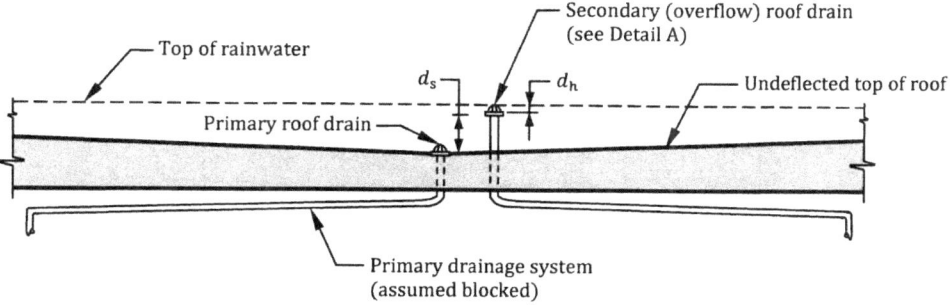

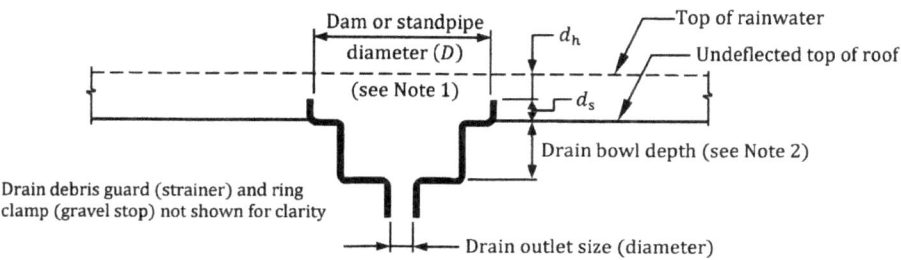

Detail A—Dam or Standpipe System

Determine d_h from ASCE Tables C8.3-1 (in.) and C8.3-2 (mm) for a given drainage system and Q.

Notes

1. For weir flow and transition flow regime designations (cells that are not shaded) in ASCE/SEI Tables C8.3-1 and C8.3-2:

   • Where the specified secondary (overflow) drain dam or standpipe diameter differs from what is provided in ASCE Tables C8.3-1 and C8.3-2, the hydraulic head can be adjusted by ASCE/SEI Eq. (C8.3-3) for a given Q:

   $$d_{h2} = (D_1/D_2)^{0.67} d_{h1} \geq 0.8 d_{h1}$$

2. For orifice flow regime designations for roof drains, as shown in the shaded cells in ASCE/SEI Tables C8.3-1 or C8.3-2:

   • Where the depth of the specified drain bowl is less than the depth of the tested drain bowl (indicated in the tables), the difference in drain bowl depth should be added to d_h from the tables to determine the design hydraulic head and total head.

   • Where the depth of the specified drain bowl is greater than that indicated in the tables, the difference in drain bowl depth can be subtracted from d_h in the tables to determine the design hydraulic head and total head. It is advisable not to use an adjusted design hydraulic head less than 80 percent of the d_h provided in the tables for a given flow rate, Q.

Figure 2.3 Determination of d_h—Roof drains.

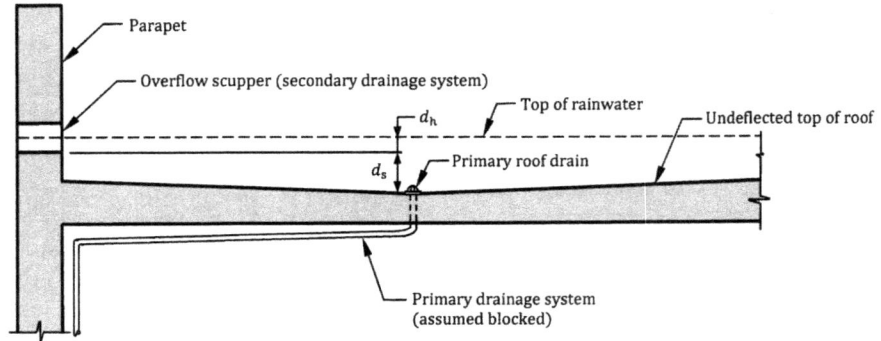

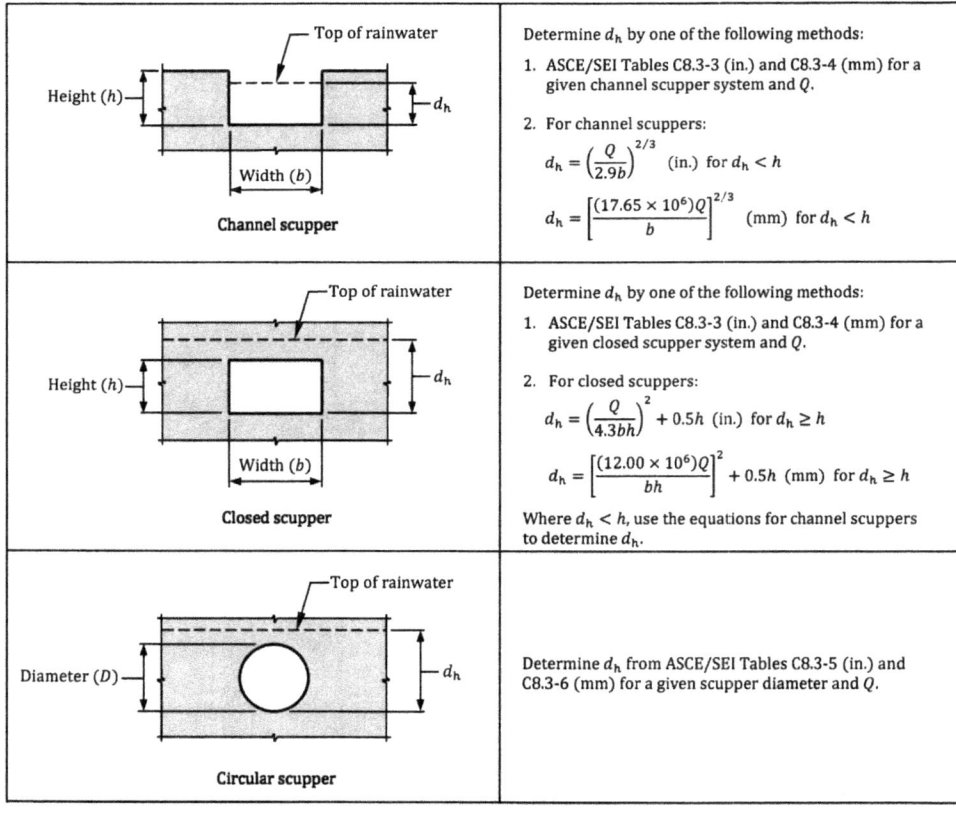

FIGURE 2.4 Determination of d_h—Scuppers.

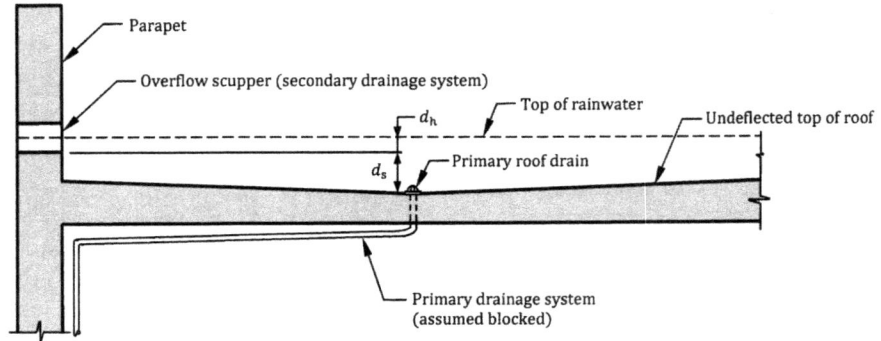

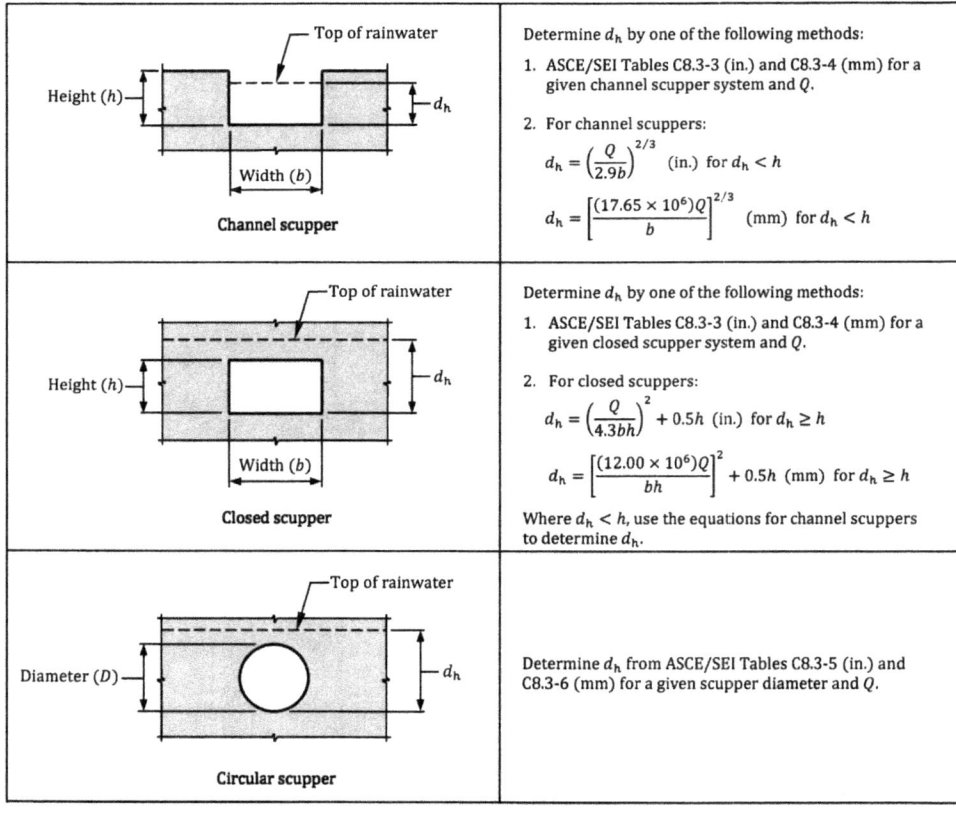

FIGURE 2.4   Determination of d_h—Scuppers.

2.8   Total Head

The total head is defined as the static head (d_s) plus the hydraulic head (d_h) associated with the design flow rate (Q) for the specified secondary drainage system. This is the depth of the rainwater used to calculate R.

2.9   Rain Load, *R*

The rain load, *R*, is calculated by IBC Equation (16-35) or ASCE/SEI Equations (8.3-1) and (8.3-1.si):

$$R = 5.2(d_s + d_h) \quad (\text{lb}/\text{ft}^2) \tag{2.3}$$

$$R = 0.0098(d_s + d_h) \quad (\text{kN}/\text{m}^2) \tag{2.4}$$

The constants in these equations are equal to the unit load of rainwater, which is the density per unit depth of rainwater:

$$\text{In Eq. (2.3):} \quad \frac{62.4 \text{ lb}/\text{ft}^3}{12 \text{ in.}/\text{ft}} = 5.2 \text{ lb}/\text{ft}^2/\text{in.}$$

$$\text{In Eq. (2.4):} \quad \frac{9.8 \text{ kN}/\text{m}^3}{1,000 \text{ mm}/\text{m}} = (0.0098 \text{ kN}/\text{m}^2)/\text{mm}$$

2.10   Ponding Instability and Ponding Loading

Where roofs do not have adequate slope or have insufficient and/or blocked drains to remove water due to rain (or melting snow), water will tend to pond in low areas, which will cause the roof structure to deflect. These low areas will subsequently attract even more water, leading to additional deflection. The structural members supporting the roof must be stiff enough so that deflections will not continually increase until instability occurs, resulting in localized failure.

An analysis for ponding instability must be performed for susceptible bays (ASCE/SEI 8.4). In general, a susceptible bay is any bay where water is impounded on the roof prior to reaching a secondary drainage system, regardless of the slope of the roof. The susceptible bays defined in ASCE/SEI 8.4 are illustrated in Figs. 2.5 through 2.8.

The roof slope limits of ¼ in. per ft (1.19 degrees) and 1 in. per ft (4.76 degrees) in ASCE/SEI 8.4 correspond to secondary members perpendicular and parallel to the

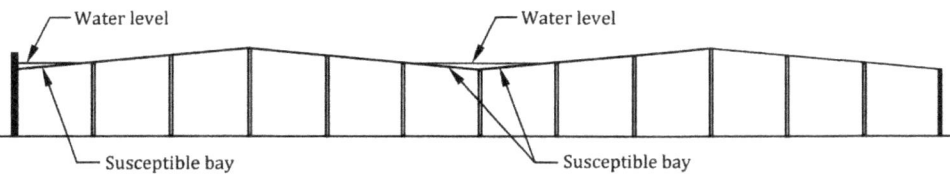

FIGURE 2.5   Example of susceptible bays for ponding evaluation.

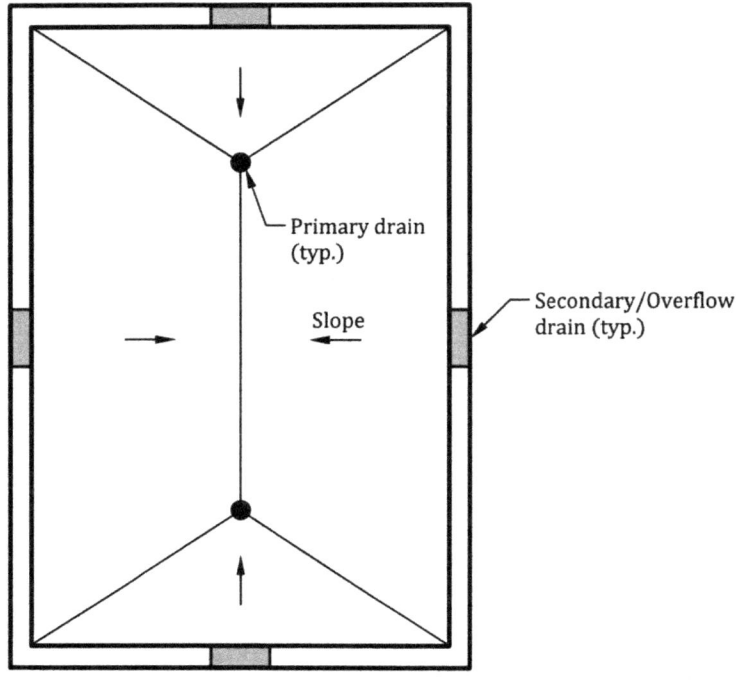

All bays are susceptible to ponding because water can accumulate (in whole or in part) when the primary drain system is blocked, and the secondary drain system is functional.

Figure 2.6 Example of a roof where all bays are susceptible to ponding.

free draining edge of a roof, respectively, and are based on a maximum deflection to span ratio of 1/240 assuming the sidewall (or primary roof member) at the free draining edge is rigid (that is, the vertical deflection of the sidewall or primary roof member is negligible compared to the deflection of the secondary and other primary members). The equations in ASCE/SEI C8.4 for the minimum roof rise, β, for a run of 1 foot form the basis of these limits. It is assumed no water is impounded in a bay where β is greater than or equal to the values determined by the following equations:

- For bays with secondary members perpendicular to the free draining edge of the roof:

$$\beta = \frac{1+(L_p/L_s)}{10} \qquad (2.5)$$

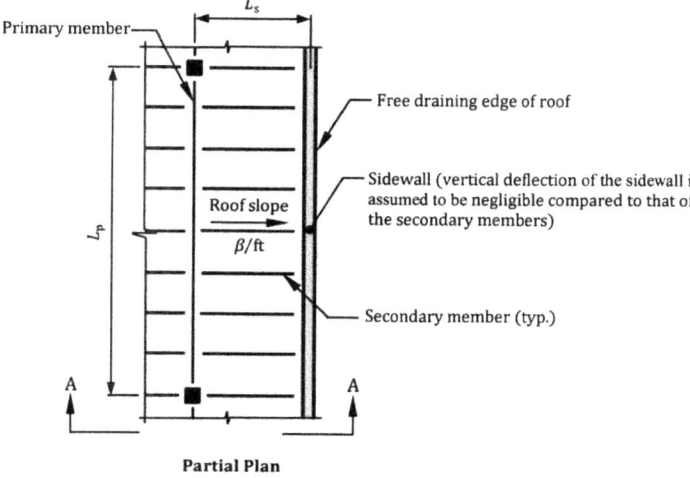

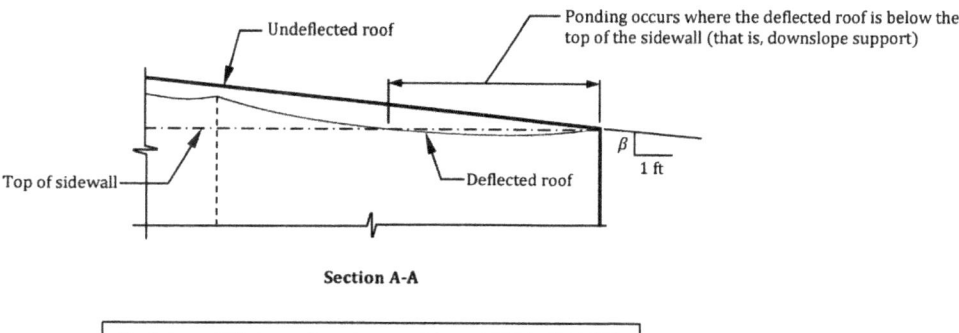

FIGURE 2.7   Susceptible bays where secondary members are perpendicular to the free draining edge of the roof.

- For bays with secondary members parallel to the free draining edge of the roof:

$$\beta = \frac{(L_s/S) + \pi}{20} \tag{2.6}$$

Values of β determined by Eq. (2.5) are given in Fig. 2.9 for primary and secondary members with spans ranging from 20 to 60 ft (6.1 to 18.3 m). Similarly, values of β determined by Eq. (2.6) are given in Fig. 2.10 for secondary beam spans and spacings ranging from 20 to 60 ft (6.1 to 18.3 m) and 3 to 12 ft (0.9 to 3.7 m), respectively.

For roofs without free draining edges, the primary and secondary structural members must have adequate flexural stiffness to avoid ponding instability.

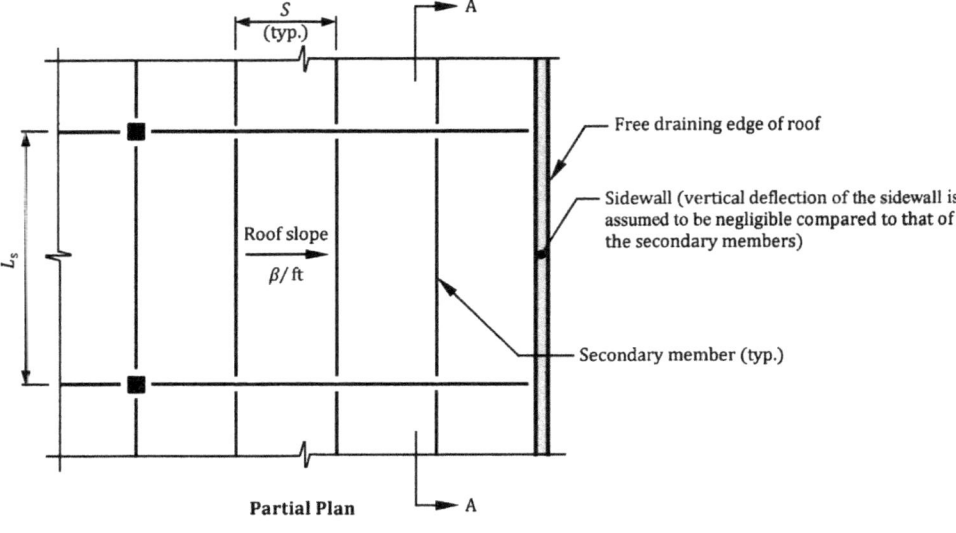

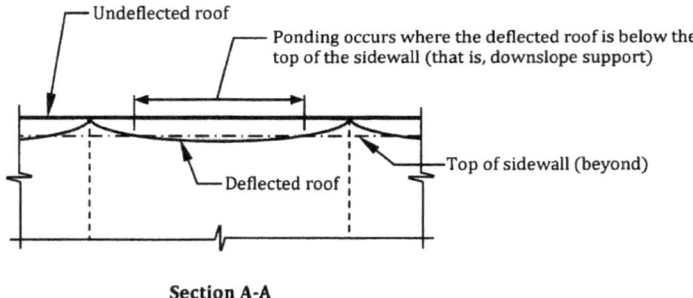

FIGURE 2.8   Susceptible bays where secondary members are parallel to the free draining edge of the roof.

In all cases, the dead load, D, and the full design rain load, R, must be used when performing a ponding analysis. A design method for roof structures considering loads from ponding is given in Ref. 6; this method accounts for the flexural rigidities of the primary and secondary structural members.

2.11   Examples

The following examples illustrate the determination of rain loads for various types of secondary drainage systems. The steps in Fig. 2.1 are used to determine R. In all examples, a rainfall intensity from a 15-minute duration/100-year return period event is used to calculate R. Examples are also provided on evaluation for ponding instability.

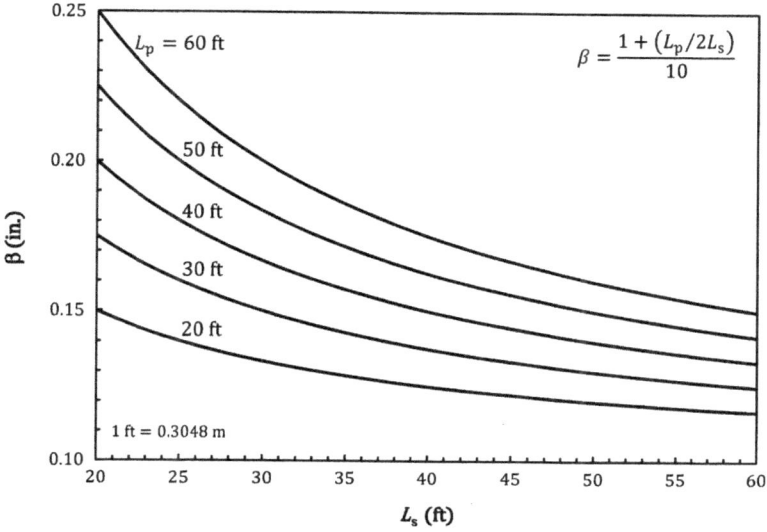

FIGURE 2.9 Roof rise, β, for susceptible bays where secondary members are perpendicular to the free draining edge of the roof.

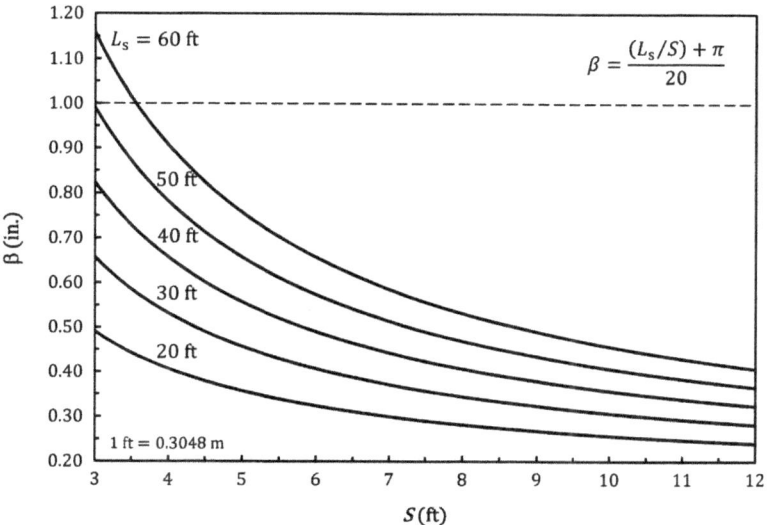

FIGURE 2.10 Roof rise, β, for susceptible bays where secondary members are parallel to the free draining edge of the roof.

2.11.1   Example 2.1—Calculation of Rain Loads for a Roof with Edge Overflow

Determine the rain load, R, on the roof in Fig. 2.11 given the design data in Table 2.2. A section at the roof edge is also shown in Fig. 2.11; the roof edge is at the same elevation on all four sides of the roof and rainwater can flow freely over the edges.

Solution

Step 1—Determine the rainfall intensity, i   Refs. 3 and 4

The rainfall intensity for a 15-minute duration/100-year return period event is equal to 6.37 in./h (161.75 mm/h) at this site.

Step 2—Calculate the flow rate, Q   Eqs. (2.1) and (2.2)

$$A = \text{Tributary area to each roof edge} = (125 \times 60)/2 = 3{,}750 \text{ ft}^2$$

$$Q = 0.0104 Ai = 0.0104 \times 3{,}750 \times 6.37 = 248.4 \text{ gal./min}$$

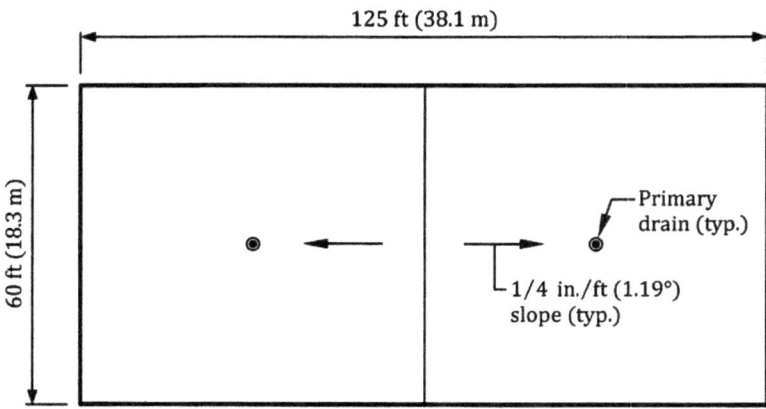

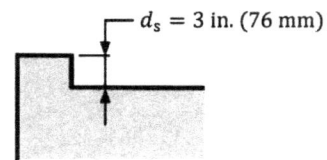

Typ. roof edge detail on all sides

Figure 2.11   Roof plan, Example 2.1.

Location	Schaumburg, IL (Latitude = 42.03°, Longitude = −88.08°)
Secondary roof drainage system	Rainwater overflow on two edges

Table 2.2   Design Data for Example 2.1

In S.I.:

$$A = \text{Tributary area to each roof edge} = (38.1 \times 18.3)/2 = 348.6 \text{ m}^2$$

$$Q = (0.278 \times 10^{-6})Ai = (0.278 \times 10^{-6}) \times 348.6 \times 161.75 = 0.0157 \text{ m}^3/\text{s}$$

Step 3—Determine the static head, d_s

From the roof section in Fig. 2.11, $d_s = 3$ in. (76 mm).

Step 4—Determine the hydraulic head, d_h Fig. 2.2

Check if the length of the roof edge, L_r, is greater than the limiting values in ASCE/SEI Equations (C8.3-2) and (C8.3-2si):

$$L_r = 60 \text{ ft} > Ai/400 = 3,750 \times 6.37/400 = 59.7 \text{ ft}$$

$$L_r = 18.3 \text{ m} > Ai/3,100 = 348.6 \times 161.75/3,100 = 18.2 \text{ m}$$

Because L_r is greater than the limiting value, $d_h = 0$.

Step 5—Calculate the rain load, R Eqs. (2.3) and (2.4)

$$R = 5.2(d_s + d_h) = 5.2 \times (3.0 + 0) = 15.6 \text{ lb/ft}^2$$

$$R = 0.0098(d_s + d_h) = 0.0098 \times (76 + 0) = 0.75 \text{ kN/m}^2$$

2.11.2 Example 2.2—Calculation of Rain Loads for a Roof with Drains (Standpipe System)

Determine the rain load, R, on the roof in Fig. 2.12 given the design data in Table 2.3. The secondary system is similar to that depicted in the upper portion of Fig. 2.3.

Solution

Step 1—Determine the rainfall intensity, i Refs. 3 and 4

The rainfall intensity for a 15-minute duration/100-year return period event is equal to 2.76 in./h (70.21 mm/h) at this site.

Step 2—Calculate the flow rate, Q Eqs. (2.1) and (2.2)

$$A = \text{Tributary area to each secondary drain} = (145 \times 85)/2 = 6,163 \text{ ft}^2$$

$$Q = 0.0104 Ai = 0.0104 \times 6,163 \times 2.76 = 176.9 \text{ gal./min}$$

In S.I.:

$$A = \text{Tributary area to each secondary drain} = (44.2 \times 25.9)/2 = 572.4 \text{ m}^2$$

$$Q = (0.278 \times 10^{-6})Ai = (0.278 \times 10^{-6}) \times 572.4 \times 70.21 = 0.0112 \text{ m}^3/\text{s}$$

Step 3—Determine the static head, d_s

From Table 2.3, $d_s = 2$ in. (51 mm).

Step 4—Determine the hydraulic head, d_h Fig. 2.3

From ASCE/SEI Table C8.3-1 for a 6-in. diameter standpipe: $d_h = 2.5$ in. for $Q = 200$ gal./min > 176.9 gal./min.

From ASCE/SEI Table C8.3-2 for a 152-mm diameter standpipe: $d_h = 64$ mm for $Q = 0.0126 \text{ m}^3/\text{s} > 0.0112 \text{ m}^3/\text{s}$.

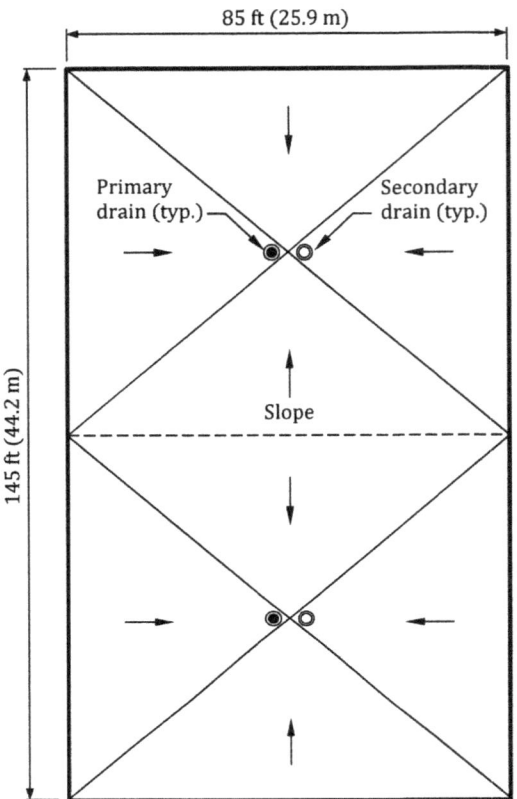

85 ft (25.9 m)

145 ft (44.2 m)

Primary
drain (typ.)

Secondary
drain (typ.)

Slope

Figure 2.12   Roof plan, Example 2.2.

Location	Reno, NV (Latitude = 39.53, Longitude = −119.82°)	
Secondary roof drainage system	Roof drains	Overflow standpipe diameter = 6 in. (152 mm)
Static head, d_s	Inlet of the overflow standpipe is set 2 in. (51 mm) above the roof surface	

Table 2.3   Design Data for Example 2.2

Step 5—Calculate the rain load, R   Eqs. (2.3) and (2.4)

$$R = 5.2(d_s + d_h) = 5.2 \times (2.0 + 2.5) = 23.4 \ \mathrm{lb/ft^2}$$
$$R = 0.0098(d_s + d_h) = 0.0098 \times (51 + 64) = 1.13 \ \mathrm{kN/m^2}$$

2.11.3   Example 2.3—Calculation of Rain Loads for a Roof with Drains (Overflow Dam System)

Determine the rain load, R, on the roof in Fig. 2.12 given the design data in Table 2.4. The secondary system is similar to Detail A in Fig. 2.3.

Location	Overland Park, KS (Latitude = 38.95°, Longitude = −94.72°)	
Secondary roof drainage system	Roof drains	Overflow dam diameter = 12.75 in. (329 mm)
		Drain outlet size = 6 in. (152 mm)
		Drain bowl depth = 2 in. (51 mm)
Static head, d_s	Overflow dam height = 2 in. (51 mm)	

TABLE 2.4 Design Data for Example 2.3

Solution

Step 1—Determine the rainfall intensity, i   Refs. 3 and 4

The rainfall intensity for a 15-minute duration/100-year return period event is equal to 7.60 in./h (193.14 mm/h) at this site.

Step 2—Calculate the flow rate, Q   Eqs. (2.1) and (2.2)

$$A = \text{Tributary area to each secondary drain} = (145 \times 85)/2 = 6,163 \text{ ft}^2$$

$$Q = 0.0104Ai = 0.0104 \times 6,163 \times 7.60 = 487.1 \text{ gal./min}$$

In S.I.:

$$A = \text{Tributary area to each secondary drain} = (44.2 \times 25.9)/2 = 572.4 \text{ m}^2$$

$$Q = (0.278 \times 10^{-6})Ai = (0.278 \times 10^{-6}) \times 572.4 \times 193.14 = 0.0307 \text{ m}^3/\text{s}$$

Step 3—Determine the static head, d_s

From Table 2.4, $d_s = 2$ in. (51 mm).

Step 4—Determine the hydraulic head, d_h   Fig. 2.3

The hydraulic head is determined by linear interpolation in ASCE/SEI Tables C8.3-1 and C8.3-2 for the given overflow dam diameter, drain outlet size, drain bowl depth, and Q:

$$d_h = 3.0 + \frac{(3.5 - 3.0) \times (487.1 - 450)}{500 - 450} = 3.4 \text{ in. at } Q = 487.1 \text{ gal./min}$$

$$d_h = 76 + \frac{(89 - 76) \times (0.0307 - 0.0284)}{0.0315 - 0.0284} = 86 \text{ mm at } Q = 0.0307 \text{ m}^3/\text{s}$$

Step 5—Calculate the rain load, R   Eqs. (2.3) and (2.4)

$$R = 5.2(d_s + d_h) = 5.2 \times (2.0 + 3.4) = 28.1 \text{ lb/ft}^2$$

$$R = 0.0098(d_s + d_h) = 0.0098 \times (51 + 86) = 1.34 \text{ kN/m}^2$$

2.11.4   Example 2.4—Calculation of Rain Loads for a Roof with Drains (Overflow Dam System with an Overflow Dam Diameter Not Given in ASCE/SEI Tables C8.3-1 and C8.3-2)

Determine the rain load, R, on the roof in Fig. 2.12 given the design data in Table 2.4 where a 10-in. (254-mm) overflow dam diameter is specified instead of the 12.75-in. (329-mm) overflow dam diameter given in Example 2.3. All other data are the same.

Solution

Step 1—Determine the rainfall intensity, i Refs. 3 and 4

The rainfall intensity for a 15-minute duration/100-year return period event is equal to 7.60 in./h (193.14 mm/h) at this site.

Step 2—Calculate the flow rate, Q Eqs. (2.1) and (2.2)

$$A = \text{Tributary area to each secondary drain} = (145 \times 85)/2 = 6,163 \text{ ft}^2$$

$$Q = 0.0104\,Ai = 0.0104 \times 6,163 \times 7.60 = 487.1 \text{ gal./min}$$

In S.I.:

$$A = \text{Tributary area to each secondary drain} = (44.2 \times 25.9)/2 = 572.4 \text{ m}^2$$

$$Q = (0.278 \times 10^{-6})Ai = (0.278 \times 10^{-6}) \times 572.4 \times 193.14 = 0.0307 \text{ m}^3/\text{s}$$

Step 3—Determine the static head, d_s

From Table 2.4, $d_s = 2$ in. (51 mm).

Step 4—Calculate the hydraulic head, d_{h2}, for the specified secondary drain Fig. 2.3, Note 1

For weir flow and transition flow regime designations (cells not shaded) in ASCE/SEI Tables C8.3-1 and C8.3-2, calculate d_{h2} by ASCE/SEI Equation (C8.3-3):

$$d_{h2} = (D_1/D_2)^{0.67} d_{h1}$$

From Step 4 in Example 2.3, $d_{h1} = 3.4$ in. (86 mm) for $Q = 487.1$ gal./min (0.0307 m^3/s) and $D_1 = 12.75$ in. (329 mm).

$$D_2 = 10.0 \text{ in. } (254 \text{ mm})$$

Therefore,

$$d_{h2} = (12.75/10.0)^{0.67} \times 3.4 = 4.0 \text{ in.}$$

$$d_{h2} = (329/254)^{0.67} \times 86 = 102 \text{ mm}$$

Step 5—Calculate the rain load, R Eqs. (2.3) and (2.4)

$$R = 5.2(d_s + d_h) = 5.2 \times (2.0 + 4.0) = 31.2 \text{ lb/ft}^2$$

$$R = 0.0098(d_s + d_h) = 0.0098 \times (51 + 102) = 1.50 \text{ kN/m}^2$$

2.11.5 Example 2.5—Calculation of Rain Loads for a Roof with Drains (Overflow Dam System) and an Adjacent Wall Diverting Rainwater onto the Roof

Determine the rain load, R, on the lower roof of the building in Fig. 2.13 given the design data in Table 2.5. The secondary system is similar to Detail A in Fig. 2.3.

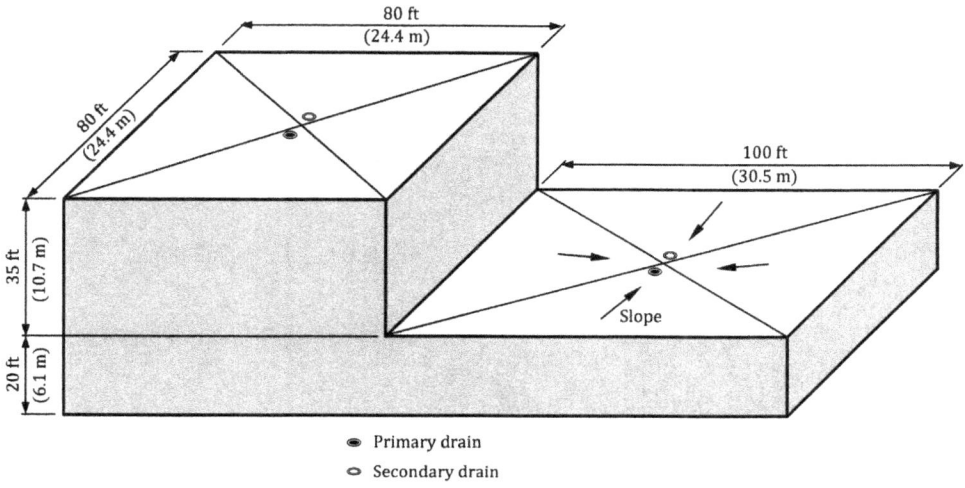

● Primary drain
○ Secondary drain

Figure 2.13   Building in Example 2.5.

Location	San Diego, CA (Latitude = 32.72°, Longitude = −117.14°)	
Secondary roof drainage system	Roof drains	Overflow dam diameter = 8 in. (203 mm)
		Drain outlet size = 4 in. (102 mm)
		Drain bowl depth = 2 in. (51 mm)
Static head, d_s	Overflow dam height = 2 in. (51 mm)	

Table 2.5   Design Data for Example 2.5

Solution

Step 1—Determine the rainfall intensity, i   Refs. 3 and 4

The rainfall intensity for a 15-minute duration/100-year return period event is equal to 2.68 in./h (68.07 mm/h) at this site.

Step 2—Calculate the flow rate, Q   Eqs. (2.1) and (2.2)

The adjacent wall diverts rainwater onto the lower roof, so A is equal to the area of the lower roof plus one-half of the wall area:

$$A = (100 \times 80) + (0.5 \times 35 \times 80) = 9,400 \text{ ft}^2$$

$$Q = 0.0104 Ai = 0.0104 \times 9,400 \times 2.68 = 262.0 \text{ gal./min}$$

In S.I.:

$$A = (30.5 \times 24.4) + (0.5 \times 10.7 \times 24.4) = 874.7 \text{ m}^2$$

$$Q = (0.278 \times 10^{-6})Ai = (0.278 \times 10^{-6}) \times 874.7 \times 68.07 = 0.0166 \text{ m}^3/\text{s}$$

Step 3—Determine the static head, d_s

From Table 2.5, $d_s = 2$ in. (51 mm).

Step 4—Determine the hydraulic head, d_h   Fig. 2.3

The hydraulic head is determined by linear interpolation in ASCE/SEI Tables C8.3-1 and C8.3-2 for the given overflow dam diameter, drain outlet size, drain bowl depth, and Q:

$$d_h = 2.5 + \frac{(3.0 - 2.5) \times (262 - 250)}{300 - 250} = 2.6 \text{ in. at } Q = 262.0 \text{ gal./min}$$

$$d_h = 64 + \frac{(76 - 64) \times (0.0166 - 0.0158)}{0.0189 - 0.0158} = 67 \text{ mm at } Q = 0.0166 \text{ m}^3/\text{s}$$

Step 5—Calculate the rain load, R   Eqs. (2.3) and (2.4)

$$R = 5.2(d_s + d_h) = 5.2 \times (2.0 + 2.6) = 23.9 \text{ lb/ft}^2$$

$$R = 0.0098(d_s + d_h) = 0.0098 \times (51 + 67) = 1.16 \text{ kN/m}^2$$

2.11.6 Example 2.6—Calculation of Rain Loads for a Roof with Drains (Overflow Dam System with a Drain Bowl Depth Not Given in ASCE/SEI Tables C8.3-1 and C8.3-2) and an Adjacent Wall Diverting Rainwater onto the Roof

Determine the rain load, R, on the lower roof of the building in Fig. 2.13 given the design data in Table 2.5 where a 1.5-in. (38-mm) drain bowl depth is specified instead of the 2.0-in. (51-mm) drain bowl depth given in Example 2.3. All other data are the same.

Solution

Step 1—Determine the rainfall intensity, i   Refs. 3 and 4

The rainfall intensity for a 15-minute duration/100-year return period event is equal to 2.68 in./h (68.07 mm/h) at this site.

Step 2—Calculate the flow rate, Q   Eqs. (2.1) and (2.2)

The adjacent wall diverts rainwater onto the lower roof, so A is equal to the area of the lower roof plus one-half of the wall area:

$$A = (100 \times 80) + (0.5 \times 35 \times 80) = 9,400 \text{ ft}^2$$

$$Q = 0.0104 Ai = 0.0104 \times 9,400 \times 2.68 = 262.0 \text{ gal./min}$$

In S.I.:

$$A = (30.5 \times 24.4) + (0.5 \times 10.7 \times 24.4) = 874.7 \text{ m}^2$$

$$Q = (0.278 \times 10^{-6}) Ai = (0.278 \times 10^{-6}) \times 874.7 \times 68.07 = 0.0166 \text{ m}^3/\text{s}$$

Step 3—Determine the static head, d_s

From Table 2.5, $d_s = 2$ in. (51 mm).

Step 4—Adjust the hydraulic head, d_h, for the specified drain bowl depth   Fig. 2.3, Note 2

The specified drain bowl depth is less than depth of the tested drain bowl and the flow regime is orifice flow (shaded portions of ASCE/SEI Tables C8.3-1 and C8.3-2). Therefore, the adjusted hydraulic head is equal to d_h from Example 2.5 plus the difference in drain bowl depths:

$$d_h = 2.6 + (2.0 - 1.5) = 3.1 \text{ in.}$$

$$d_h = 67 + (51 - 38) = 80 \text{ mm}$$

Step 5—Calculate the rain load, R   Eqs. (2.3) and (2.4)

$$R = 5.2(d_s + d_h) = 5.2 \times (2.0 + 3.1) = 26.5 \text{ lb/ft}^2$$

$$R = 0.0098(d_s + d_h) = 0.0098 \times (51 + 80) = 1.28 \text{ kN/m}^2$$

2.11.7   Example 2.7—Calculation of Rain Loads for a Roof with Channel Scuppers

Determine the rain load, R, on the roof in Fig. 2.14 given the design data in Table 2.6 using four channel scuppers.

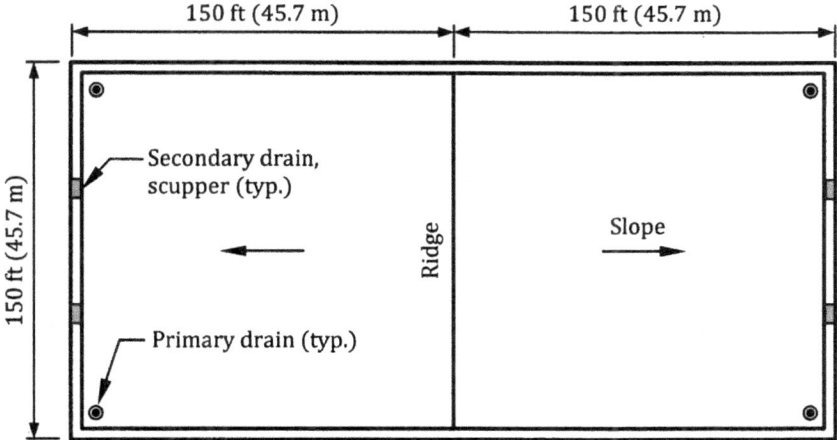

FIGURE 2.14   Roof plan, Example 2.7.

Location	Jacksonville, FL (Latitude = 30.35, Longitude = −81.73°)	
Secondary roof drainage system	Channel scuppers	Scupper width (b) = 24 in. (610 mm)
		Height (h) = 12 in. (305 mm)
Static head, d_s	Inlet of the scuppers is set 2 in. (51 mm) above the roof surface	

TABLE 2.6   Design Data for Example 2.7

Solution

Step 1—Determine the rainfall intensity, i   Refs. 3 and 4

The rainfall intensity for a 15-minute duration/100-year return period event is equal to 8.37 in./h (212.65 mm/h) at this site.

Step 2—Calculate the flow rate, Q   Eqs. (2.1) and (2.2)

$$A = \text{Tributary area to each secondary scupper} = (300 \times 150)/4 = 11,250 \text{ ft}^2$$

$$Q = 0.0104 Ai = 0.0104 \times 11,250 \times 8.37 = 979.3 \text{ gal./min}$$

In S.I.:

$$A = \text{Tributary area to each secondary scupper} = (91.4 \times 45.7)/4 = 1,044.3 \text{ m}^2$$

$$Q = (0.278 \times 10^{-6}) Ai = (0.278 \times 10^{-6}) \times 1,044.3 \times 212.65 = 0.0617 \text{ m}^3/\text{s}$$

Step 3—Determine the static head, d_s

From Table 2.6, $d_s = 2$ in. (51 mm).

Step 4—Determine the hydraulic head, d_h   Fig. 2.4

For channel scuppers where $d_h < h$:

$$d_h = \left(\frac{Q}{2.9b}\right)^{2/3} = \left(\frac{979.3}{2.9 \times 24.0}\right)^{2/3} = 5.8 \text{ in.} < h = 12.0 \text{ in.}$$

$$d_h = \left[\frac{(17.65 \times 10^6)Q}{b}\right]^{2/3} = \left[\frac{(17.65 \times 10^6) \times 0.0617}{610}\right]^{2/3} = 147 \text{ mm} < h = 305 \text{ mm}$$

Alternatively, determine d_h by linear interpolation in ASCE/SEI Tables C8.3-3 and C8.3-4 for the given scupper width and Q:

$$d_h = 5.0 + \frac{(7-5) \times (979.3 - 776)}{1,284 - 776} = 5.8 \text{ in. at } Q = 979.3 \text{ gal./min}$$

$$d_h = 127 + \frac{(178 - 127) \times (0.0617 - 0.0490)}{0.0810 - 0.0490} = 147 \text{ mm at } Q = 0.0617 \text{ m}^3/\text{s}$$

Step 5—Calculate the rain load, R   Eqs. (2.3) and (2.4)

$$R = 5.2(d_s + d_h) = 5.2 \times (2.0 + 5.8) = 40.6 \text{ lb/ft}^2$$

$$R = 0.0098(d_s + d_h) = 0.0098 \times (51 + 147) = 1.94 \text{ kN/m}^2$$

2.11.8   Example 2.8—Calculation of Rain Loads for a Roof with Closed Scuppers

Determine the rain load, R, on the roof in Fig. 2.14 given the design data in Table 2.6 using four 24-in. (610-mm) wide by 4-in. (102-mm) high closed scuppers instead of the channel scuppers. All other data are the same.

Solution

Step 1—Determine the rainfall intensity, i   Refs. 3 and 4

The rainfall intensity for a 15-minute duration/100-year return period event is equal to 8.37 in./h (212.65 mm/h) at this site.

Step 2—Calculate the flow rate, Q   Eqs. (2.1) and (2.2)

A = Tributary area to each secondary scupper = $(300 \times 150)/4 = 11,250$ ft^2

$Q = 0.0104Ai = 0.0104 \times 11,250 \times 8.37 = 979.3$ gal./min

In S.I.:

A = Tributary area to each secondary scupper = $(91.4 \times 45.7)/4 = 1,044.3$ m^2

$Q = (0.278 \times 10^{-6})Ai = (0.278 \times 10^{-6}) \times 1,044.3 \times 212.65 = 0.0617$ m^3/s

Step 3—Determine the static head, d_s

From Table 2.6, $d_s = 2$ in. (51 mm).

Step 4—Determine the hydraulic head, d_h   Fig. 2.4

For closed scuppers where $d_h > h$:

$$d_h = \left(\frac{Q}{4.3bh}\right)^2 + 0.5h = \left(\frac{979.3}{4.3 \times 24.0 \times 4.0}\right)^2 + (0.5 \times 4.0) = 7.6 \text{ in.} > h = 4.0 \text{ in.}$$

$$d_h = \left[\frac{(12.00 \times 10^6)Q}{bh}\right]^2 + 0.5h = \left[\frac{(12.00 \times 10^6) \times 0.0617}{610 \times 102}\right]^2$$
$$+ (0.5 \times 102) = 193 \text{ mm} > h = 102 \text{ mm}$$

Alternatively, determine d_h by linear interpolation in ASCE/SEI Tables C8.3-3 and C8.3-4 for the given scupper size and Q:

$$d_h = 7.0 + \frac{(8-7) \times (979.3 - 924)}{1,012 - 924} = 7.6 \text{ in. at } Q = 979.3 \text{ gal./min}$$

$$d_h = 178 + \frac{(203 - 178) \times (0.0617 - 0.0583)}{0.0638 - 0.0583} = 194 \text{ mm at } Q = 0.0617 \text{ m}^3/\text{s}$$

Step 5—Calculate the rain load, R   Eqs. (2.3) and (2.4)

$$R = 5.2(d_s + d_h) = 5.2 \times (2.0 + 7.6) = 49.9 \text{ lb/ft}^2$$

$$R = 0.0098(d_s + d_h) = 0.0098 \times (51 + 194) = 2.40 \text{ kN/m}^2$$

2.11.9 Example 2.9—Calculation of Rain Loads for a Roof with Circular Scuppers

Determine the rain load, R, on the roof in Fig. 2.14 given the design data in Table 2.6 using eight 12-in. (305-mm) diameter circular scuppers (four scuppers each face) instead of the channel scuppers. All other data are the same.

Solution

Step 1—Determine the rainfall intensity, i   Refs. 3 and 4

The rainfall intensity for a 15-minute duration/100-year return period event is equal to 8.37 in./h (212.65 mm/h) at this site.

Step 2—Calculate the flow rate, Q   Eqs. (2.1) and (2.2)

$$A = \text{Tributary area to each secondary scupper} = (300 \times 150)/8 = 5,625 \text{ ft}^2$$

$$Q = 0.0104 Ai = 0.0104 \times 5,625 \times 8.37 = 489.7 \text{ gal./min}$$

In S.I.:

$$A = \text{Tributary area to each secondary scupper} = (91.4 \times 45.7)/8 = 522.1 \text{ m}^2$$

$$Q = (0.278 \times 10^{-6})Ai = (0.278 \times 10^{-6}) \times 522.1 \times 212.65 = 0.0309 \text{ m}^3/\text{s}$$

Step 3—Determine the static head, d_s

From Table 2.6, $d_s = 2$ in. (51 mm).

Step 4—Determine the hydraulic head, d_h   Fig. 2.4

The hydraulic head is determined by linear interpolation in ASCE/SEI Tables C8.3-5 and C8.3-6 for the given circular scupper diameter and Q:

$$d_h = 7.0 + \frac{(8-7) \times (489.7 - 410)}{510 - 410} = 7.8 \text{ in. at } Q = 489.7 \text{ gal./min}$$

$$d_h = 178 + \frac{(203 - 178) \times (0.0309 - 0.0259)}{0.0322 - 0.0259} = 198 \text{ mm at } Q = 0.0309 \text{ m}^3/\text{s}$$

Step 5—Calculate the rain load, R   Eqs. (2.3) and (2.4)

$$R = 5.2(d_s + d_h) = 5.2 \times (2.0 + 7.8) = 51.0 \text{ lb/ft}^2$$

$$R = 0.0098(d_s + d_h) = 0.0098 \times (51 + 198) = 2.44 \text{ kN/m}^2$$

2.11.10   Example 2.10—Determination of Minimum Roof Slope to Avoid Ponding Instability

Rainwater is free to drain over the roof edge for the framing system shown in Fig. 2.15. Assuming the sidewall is rigid, determine the roof slope to avoid ponding instability.

Solution

The secondary members are perpendicular to the free draining edge of the roof. Therefore, determine the minimum roof rise, β, by Eq. (2.5):

$$\beta = \frac{1 + (L_p/L_s)}{10} = \frac{1 + [25.0/(2 \times 30.0)]}{10} = 0.14 \text{ in. (4 mm)}$$

Provide a roof slope of at least 0.25 in./ft (1.19 degrees) to avoid ponding instability.

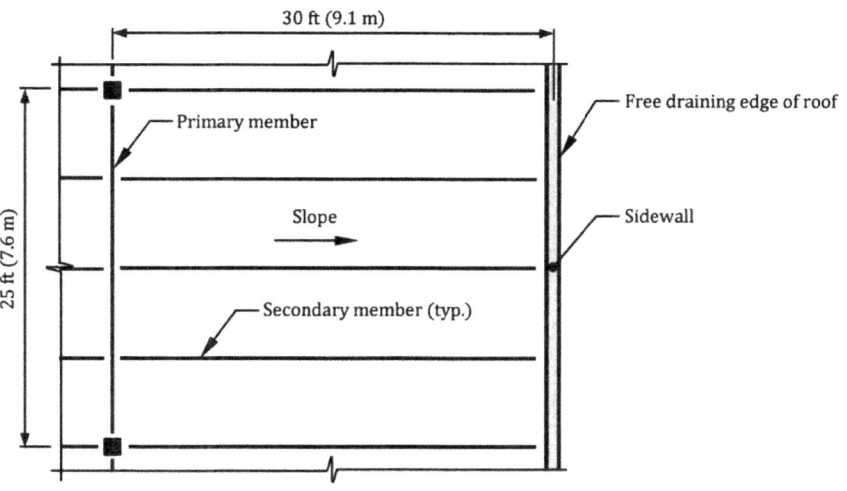

Figure 2.15 Roof framing plan, Example 2.10.

2.11.11   Example 2.11—Determination of Secondary Member Spacing to Avoid Ponding Instability

Rainwater is free to drain over the roof edge for the framing system shown in Fig. 2.16. Assuming the sidewall is rigid, determine the spacing of the secondary members to avoid ponding instability.

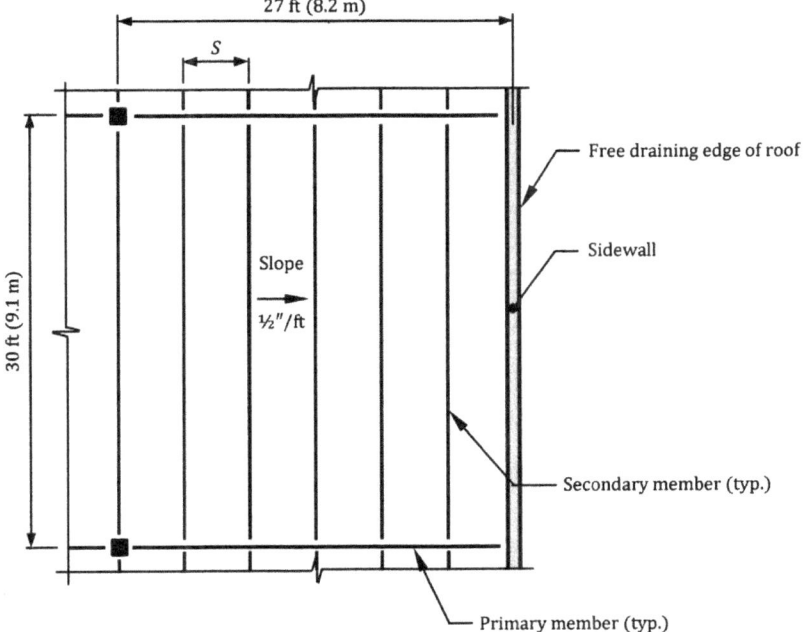

Figure 2.16 Roof framing plan, Example 2.11.

Solution

The secondary members are parallel to the free draining edge of the roof. Therefore, determine the spacing, S, by Eq. (2.6):

$$\beta = \frac{(L_s/S) + \pi}{20}$$

Solving for S:

$$S = \frac{L_s}{20\beta - \pi} = \frac{30.0}{(20 \times 0.5) - \pi} = 4.4 \text{ ft (1.3 m)}$$

The secondary members are to be spaced at 4 ft-6 in. (1.4 m) on center to avoid ponding instability.

CHAPTER 3

Snow Loads

3.1 Overview

This chapter contains methods to calculate design snow loads in accordance with IBC 1608 and ASCE/SEI Chapter 7. Structural members supporting roofs, balconies, canopies, and similar structures must be designed for the effects of snow loads in areas where snowfall can occur.

3.2 Notation

C_e = exposure factored determined from ASCE/SEI Table 7.3-1

C_s = slope factor determined from ASCE/SEI Figure 7.4-1

$C_{s|eave}$ = slope factor determined from ASCE/SEI Figure 7.4-1 at the eaves of the roof

$C_{s|30}$ = slope factor determined from ASCE/SEI Figure 7.4-1 at the location where the roof slope is equal to 30 degrees

C_t = thermal factor determined from ASCE/SEI Table 7.3-2

D = diameter of pipe or width of cable tray, in. (mm)

h = vertical separation distance between the edge of a higher roof including any parapet and the edge of a lower adjacent roof excluding any parapet, ft (m)

h_b = height of balanced snow load = p_s/γ, ft (m)

h_c = clear height from top of balanced snow load to (1) closet point on adjacent upper roof, (2) top of parapet, or (3) top of a projection on the roof, ft (m)

h_d = height of snow drift, ft (m)

$h_{d,leeward}$ = height of leeward snow drift, ft (m)

$h_{d,windward}$ = height of windward snow drift, ft (m)

h_{d1}, h_{d2} = height of snow drift where two intersecting snow drifts can form, ft (m)

h_o = height of obstruction above the surface of the roof, ft (m)

h_r = vertical distance from a roof valley to a roof ridge, ft (m)

h_{step} = vertical distance from the lower roof to the upper roof; height of a projection; or, height of a parapet wall, ft (m)

I_s = importance factor prescribed in ASCE/SEI 7.3.3

ℓ_u = length of the roof upwind of the drift, ft (m)

ℓ_{upper} = length of the upper roof (fetch distance) in the direction of the wind, ft (m)

ℓ_{lower} = length of the lower roof (fetch distance) in the direction of the wind, ft (m)

n = number of spans in a continuous beam system

p_d = maximum intensity of drift surcharge load, lb/ft² (kN/m²)

p_f = snow load on flat roofs, lb/ft² (kN/m²)

$p_{f(heated)}$ = flat roof snow load on heated portions of roofs, lb/ft² (kN/m²)

$p_{f(unheated)}$ = flat roof snow load on unheated portions of roofs, lb/ft² (kN/m²)

p_g = ground snow load, lb/ft² (kN/m²)

p_m = minimum snow load for low-slope roofs, lb/ft² (kN/m²)

p_s = sloped roof (balanced) snow load, lb/ft² (kN/m²)

$p_{sliding}$ = sliding snow load, lb/ft (kN/m) or lb/ft² (kN/m²)

p_{total} = total snow load, lb/ft² (kN/m²)

s = horizontal separation distance between the edges of two adjacent buildings, ft (m)

S = roof slope run for a rise of one

S_p = clear spacing between multiple pipes or cable trays at the same elevation, ft (m)

w = width of snow drift, ft (m)

w_1, w_2 = width of snow drift where two intersecting snow drifts can form, ft (m)

W = horizontal distance from eave to ridge, ft (m)

γ = snow density, lb/ft³ (kN/m³)

θ = roof slope on the leeward side, degrees

3.3 Procedure to Determine Design Snow Loads

A step-by-step procedure to determine design snow loads is given in Fig. 3.1 (see ASCE/SEI Chapter C7). The tables and figures referenced in Fig. 3.1 contain additional information needed to calculate snow loads; some of the figures contain flowcharts.

3.4 Ground Snow Load, p_g

Ground snow load, p_g, in pounds per square foot (lb/ft²) can be obtained from the figures and tables in IBC 1608.2 and ASCE/SEI 7.2 (see Table 3.1; to convert lb/ft² to kN/m², multiply by 0.0479).

Ground snow loads are given in IBC Figure 1608.2 for all contiguous states and in IBC Table 1608.2 and ASCE/SEI Table 7.2-1 for select locations in Alaska. The map in ASCE/SEI Figure 7.2-1 contains the same ground snow loads as in IBC Figure 1608.2 except for the states identified in Table 3.1; for those states, ground snow loads for select

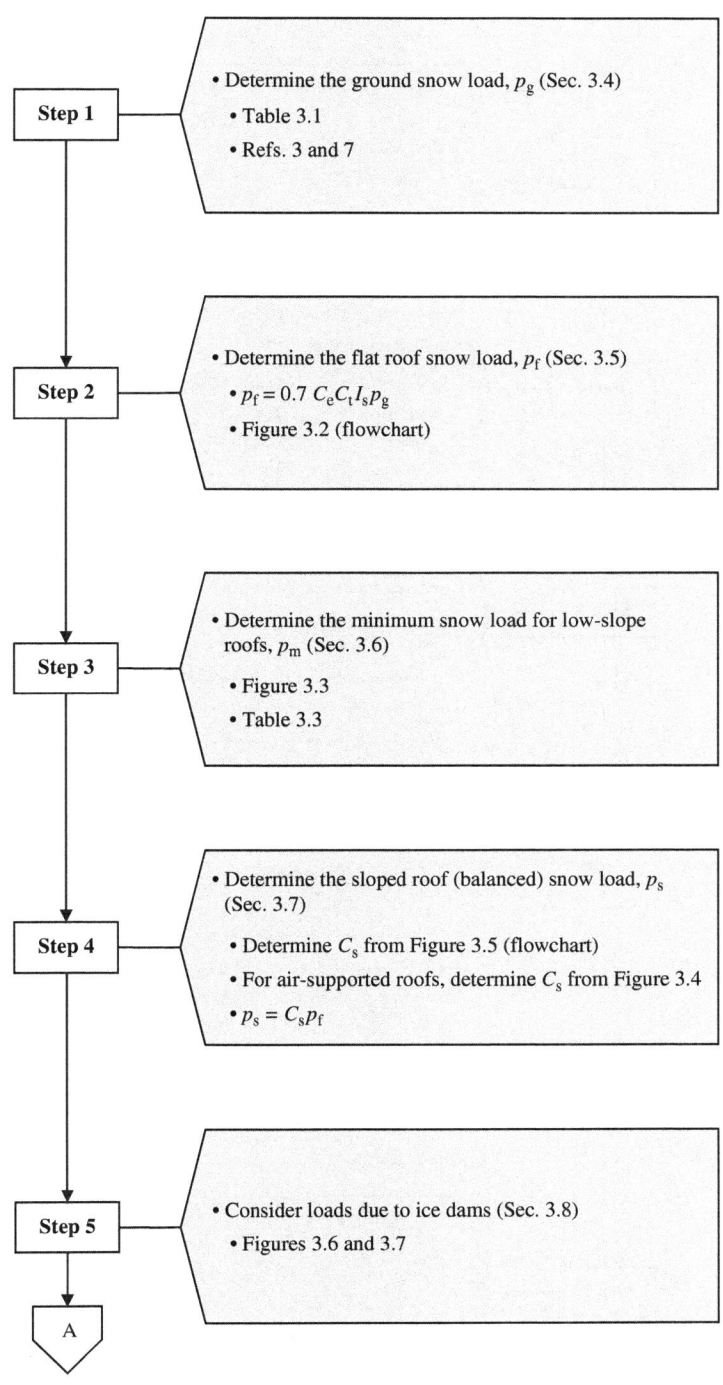

FIGURE 3.1 Procedure to determine design snow loads.

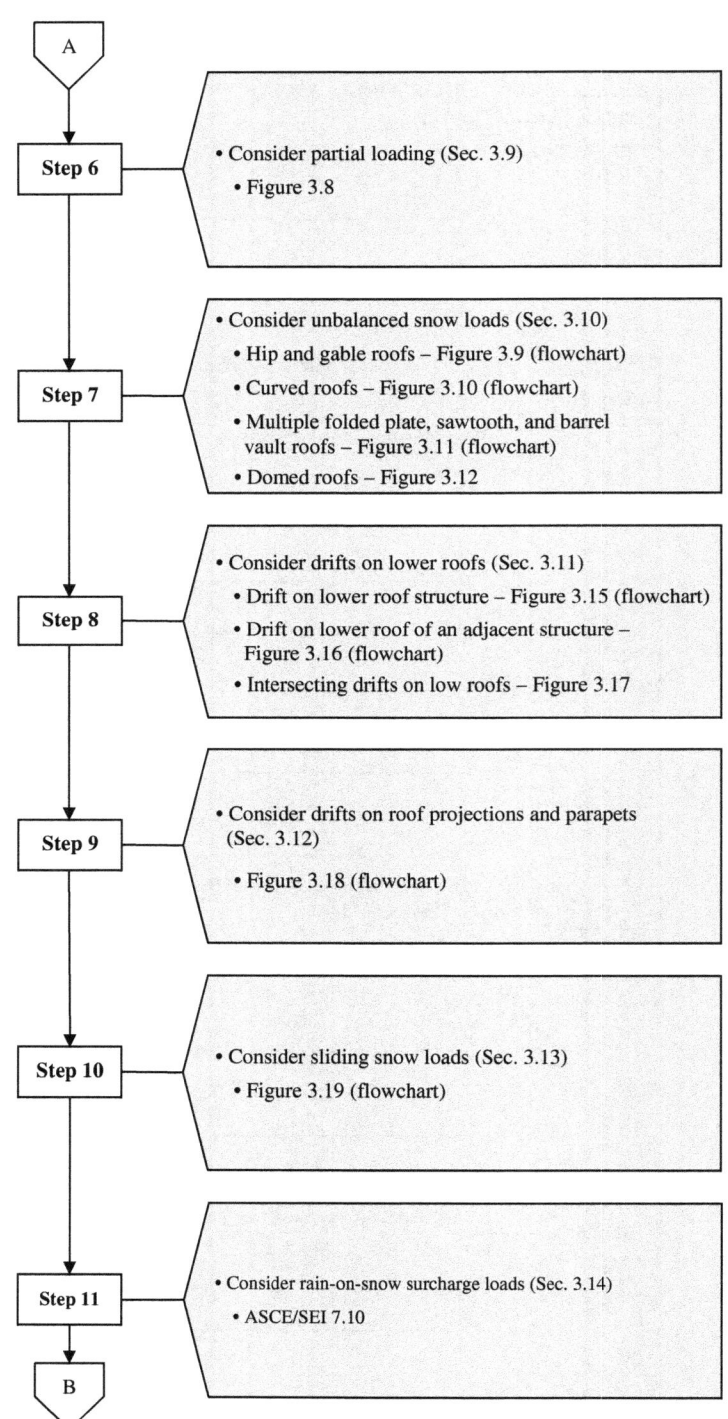

A

Step 6

- Consider partial loading (Sec. 3.9)
  - Figure 3.8

Step 7

- Consider unbalanced snow loads (Sec. 3.10)
  - Hip and gable roofs – Figure 3.9 (flowchart)
  - Curved roofs – Figure 3.10 (flowchart)
  - Multiple folded plate, sawtooth, and barrel vault roofs – Figure 3.11 (flowchart)
  - Domed roofs – Figure 3.12

Step 8

- Consider drifts on lower roofs (Sec. 3.11)
  - Drift on lower roof structure – Figure 3.15 (flowchart)
  - Drift on lower roof of an adjacent structure – Figure 3.16 (flowchart)
  - Intersecting drifts on low roofs – Figure 3.17

Step 9

- Consider drifts on roof projections and parapets (Sec. 3.12)
  - Figure 3.18 (flowchart)

Step 10

- Consider sliding snow loads (Sec. 3.13)
  - Figure 3.19 (flowchart)

Step 11

- Consider rain-on-snow surcharge loads (Sec. 3.14)
  - ASCE/SEI 7.10

B

Figure 3.1   *(Continued)*

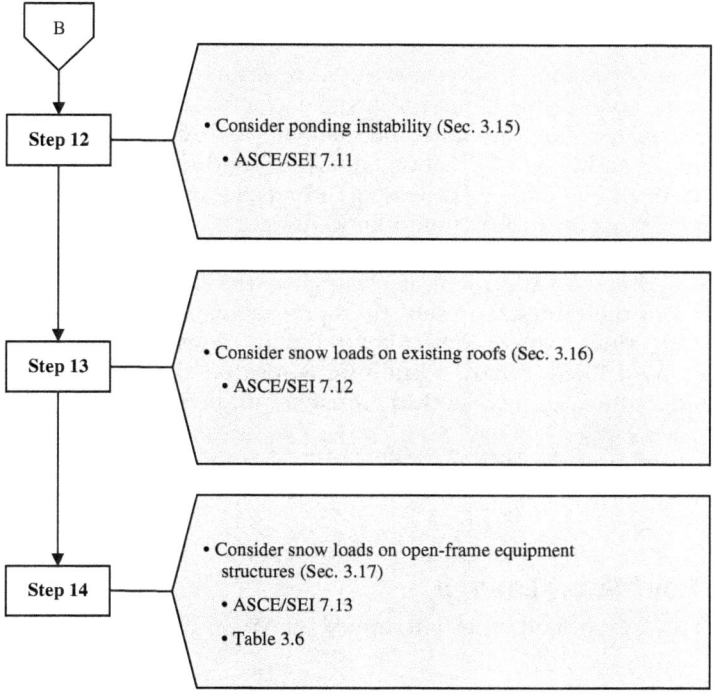

Figure 3.1 *(Continued)*

Location*	Source
Contiguous United States	IBC Figure 1608.2
	ASCE/SEI Figure 7.2-1
Alaska	IBC Table 1608.2
	ASCE/SEI Table 7.2-1
Colorado	ASCE/SEI Table 7.2-2
Idaho	ASCE/SEI Table 7.2-3
Montana	ASCE/SEI Table 7.2-4
Washington	ASCE/SEI Table 7.2-5
New Mexico	ASCE/SEI Table 7.2-6
Oregon	ASCE/SEI Table 7.2-7
New Hampshire	ASCE/SEI Table 7.2-8

*For Hawaii, ground snow loads are zero except in mountainous regions as determined by the authority having jurisdiction (IBC 1608.2 and ASCE/SEI 7.2).

Table 3.1 Sources for Determination of Ground Snow Load, p_g, in the IBC and ASCE/SEI 7

locations can be obtained from ASCE/SEI Tables 7.2-2 through 7.2-8. In all cases, the ground snow loads have been determined based on a 2-percent annual probability of being exceeded (that is, a 50-year mean recurrence interval).

In some areas of the United States, the ground snow load is too variable to allow mapping. Such regions are noted on the maps as "CS," which indicates a site-specific case study is required. Information on how to conduct a site-specific case study can be found in ASCE/SEI C7.2. It is always good practice to confirm ground snow loads with the authority having jurisdiction prior to design.

In some regions of the maps and in all the tables, ground snow loads are provided based on elevation. Numbers in parentheses in the maps and the elevations provided in the tables represent the upper elevation limits in feet for the ground snow load values to be used for sites below that elevation (to convert feet to meters, multiply by 0.3048). Where a building is located at an elevation greater than that indicated, a site-specific case study must be conducted to establish the ground snow load.

In addition to the IBC and ASCE/SEI 7, ground snow loads can be obtained by entering an address or the latitude and longitude of the site in Refs. 3 or 7.

3.5　Flat Roof Snow Load, p_f

The flat roof snow load, p_f, is determined by ASCE/SEI Equation (7.3-1):

$$p_f = 0.7C_eC_tI_sp_g \text{ (lb/ft}^2 \text{ and kN/m}^2) \tag{3.1}$$

This equation is applicable to the design of a flat roof, which is defined in ASCE/SEI 7.1.2 as a roof with a slope less than or equal to 5 degrees (see the definition of p_f in ASCE/SEI 7.1.2).

The 0.7 factor in Eq. (3.1) is a ground-to-roof conversion factor; this factor accounts for the results from research studies, which have shown that the snow load on a roof is less than that on the ground in cases where drifting is not prevalent.

The terms C_e, C_t, and I_s account for the exposure, thermal, and occupancy characteristics of the roof and building, respectively, and are determined from the ASCE/SEI 7 tables identified in Table 3.2.

The flowchart in Fig. 3.2 can be used to determine p_f.

Item	ASCE/SEI Table No.
Exposure factor, C_e	7.3-1
Thermal factor, C_t	7.3-2
Importance factor, I_s	1.5-2

TABLE 3.2　Sources for Determination of C_e, C_t, and I_s

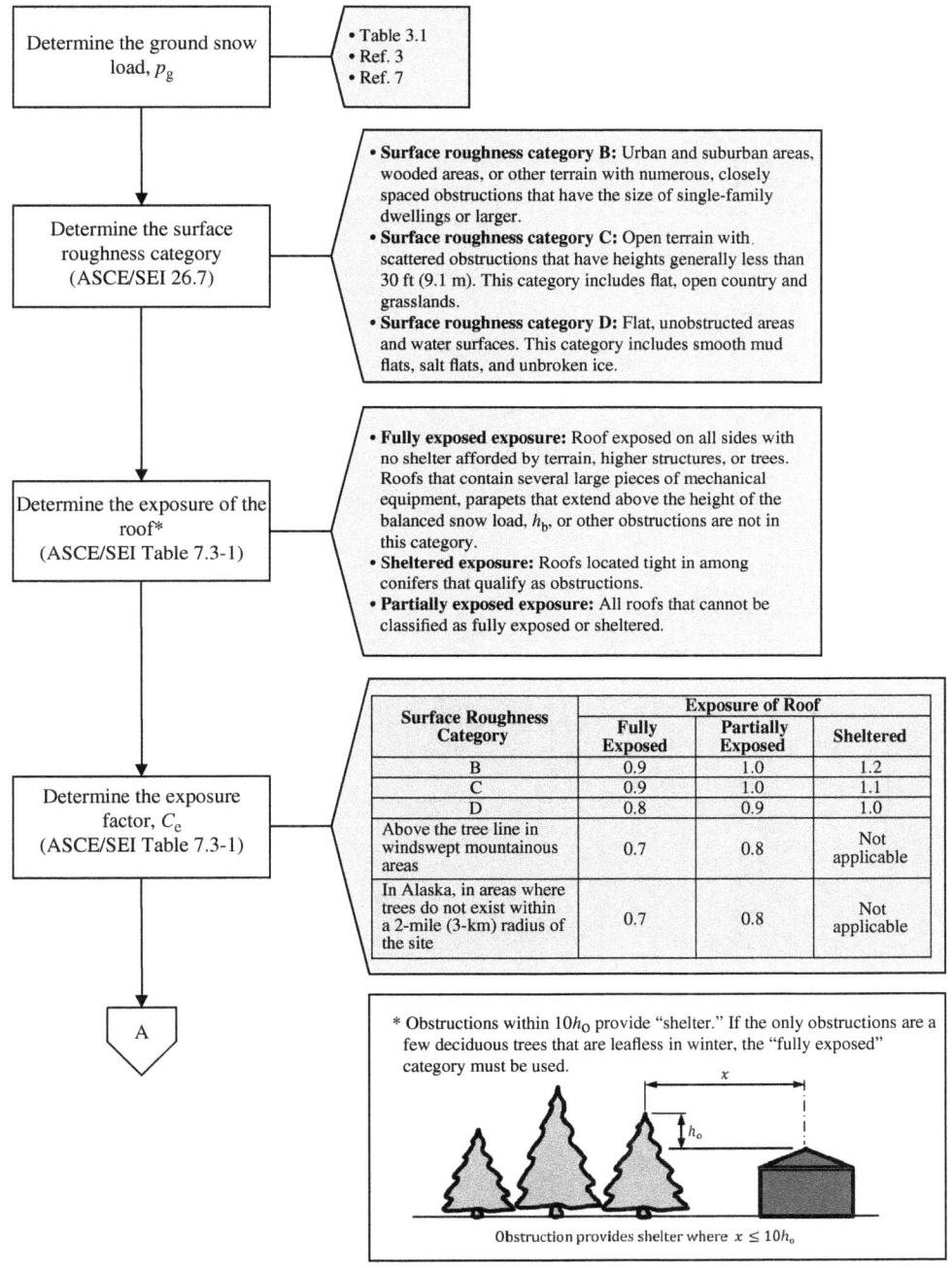

Figure 3.2 Flowchart to determine the flat roof snow load, p_f.

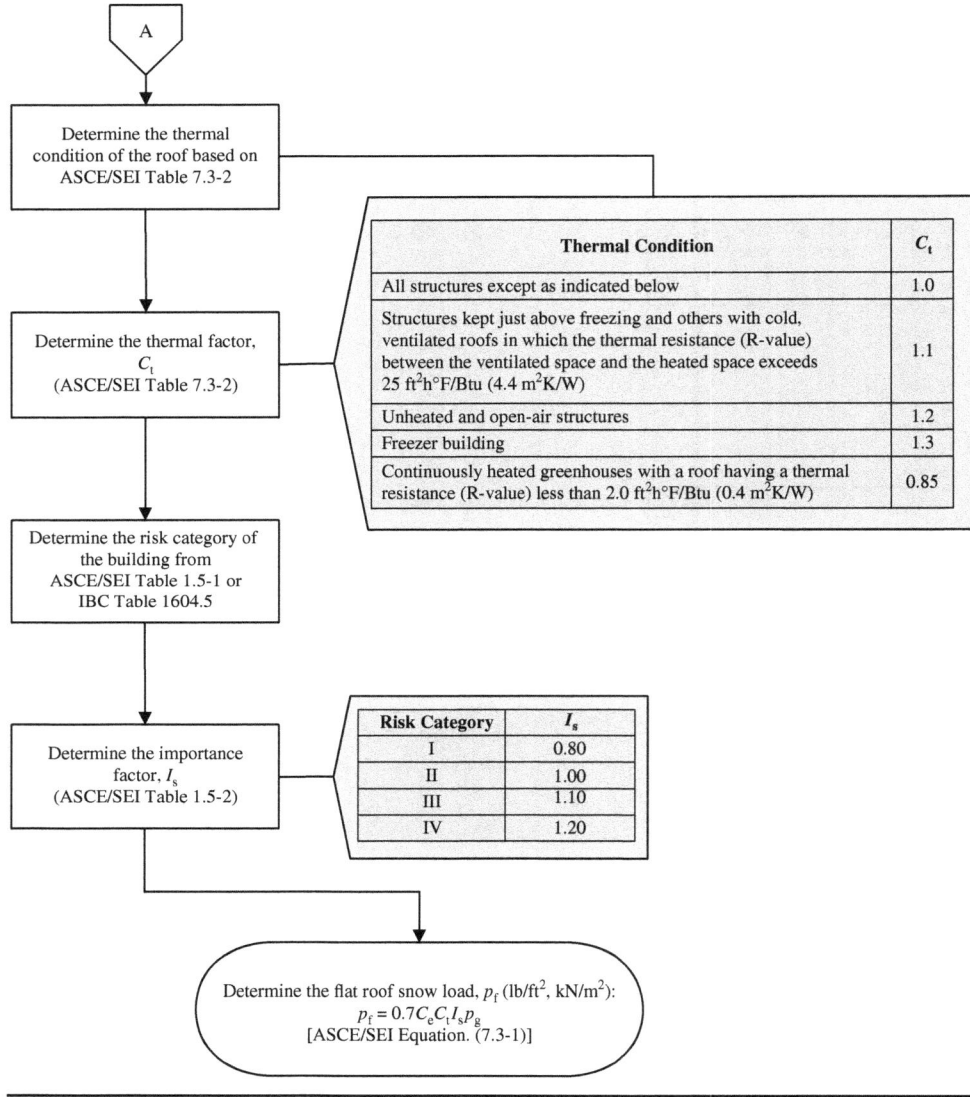

FIGURE 3.2 *(Continued)*

3.6 Minimum Snow Load for Low-Slope Roofs, p_m

Low-slope roofs are defined in ASCE/SEI 7.3.4 for three types of roof configurations: monoslope, hip or gable, and curved (see Fig. 3.3).

Minimum roof snow loads, p_m, to be applied to low-slope roofs are given in Table 3.3. This minimum snow load is a separate uniform load case; it need not be used in determining or in combination with drift, sliding, unbalanced, or partial loads.

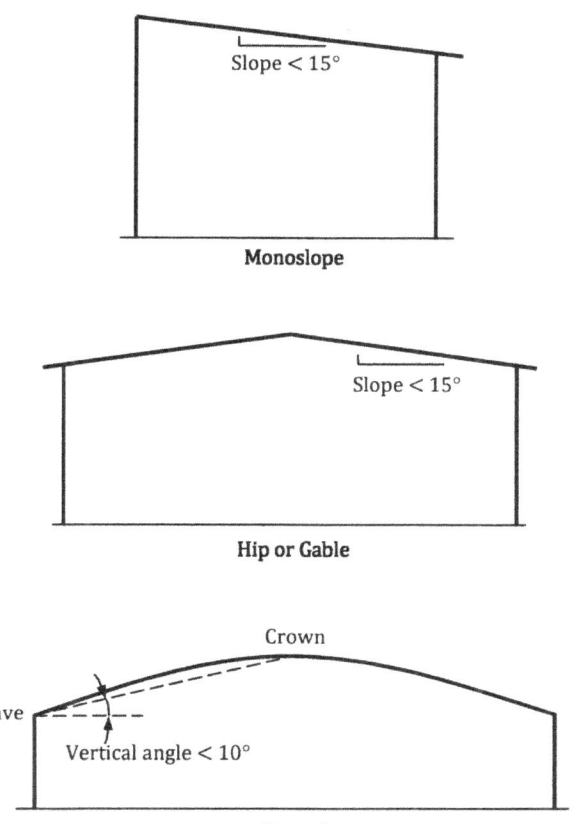

Monoslope

Hip or Gable

Curved

Figure 3.3   Low-slope roofs.

Ground Snow Load, p_g	Minimum Roof Snow Load, p_m
≤ 20 lb/ft² (0.96 kN/m²)	$I_s p_g$
> 20 lb/ft² (0.96 kN/m²)	$20 I_s$ (lb/ft²)
	$0.96 I_s$ (kN/m²)

Table 3.3   Minimum Roof Snow Loads, p_m

3.7   Sloped Roof Snow Loads, p_s

For other than flat roofs, the sloped roof (balanced) snow load, p_s, is determined by ASCE/SEI Equation (7.4-1):

$$p_s = C_s p_f \quad \text{(lb/ft² and kN/m²)} \tag{3.2}$$

It is assumed this snow load acts on the horizontal projection of the roof surface.

Slippery roof surfaces	Metal
	Slate
	Glass
	Bituminous membranes*
	Rubber membranes*
	Plastic membranes*
Not Slippery roof surfaces	Asphalt shingles
	Wood shingles
	Shakes

*For a membrane roof system to be considered slippery, the membrane must be smooth. Membranes with an embedded aggregate or a mineral granule surface are not considered to be smooth.

TABLE 3.4   Slippery and Not Slippery Roof Surfaces

The slope factor, C_s, is determined in accordance with ASCE/SEI 7.4.1 through 7.4.4 and depends on the slope and temperature of the roof, the presence or absence of obstructions, and the degree of slipperiness of the roof surface.

A warm roof is one where the thermal factor, C_t, determined in accordance with ASCE/SEI Table 7.3-2 is less than or equal to 1.0. A cold roof is one where $C_t > 1.0$.

According to ASCE/SEI 7.4, a roof surface is considered to be unobstructed if there are no (1) objects on the roof preventing the snow from sliding off (such as equipment, a snow retention device, or an ice dam) and (2) obstructions below the eaves preventing the snow from sliding to a space able to accept it all. Examples of roof surfaces considered to be slippery and not slippery are given in Table 3.4.

Graphs to determine C_s for warm and cold roofs based on whether the roof surface is an unobstructed slippery surface or not are given in ASCE/SEI Figure 7.4-1. Equations to determine C_s are given in ASCE/SEI C7.4.

For structures with air-supported roofs with vinyl coated exterior fabric (which is considered to be slippery), C_s varies linearly from 0.6 at the location where the roof slope is 30 degrees to 1.0 at the location where the slope is 5 degrees (see ASCE/SEI Figure 7.4-3). Values of C_s in ASCE/SEI Figure 7.4-3 match the values in ASCE/SEI Figure 7.4-1(a) up to a roof slope of 30 degrees for unobstructed slippery surfaces with $C_t = 1.0$. Snow loads are assumed to be zero for roof slopes greater than 30 degrees. A snow load diagram for structures with air-supported roofs is given in Fig. 3.4.

The flowchart in Fig. 3.5 can be used to determine C_s.

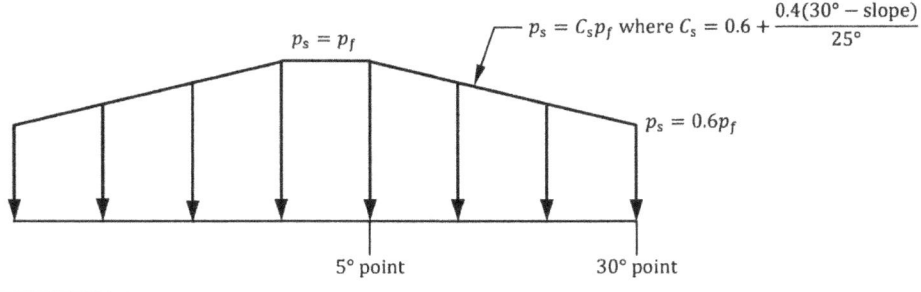

FIGURE 3.4   Snow load diagram for structures with air-supported roofs.

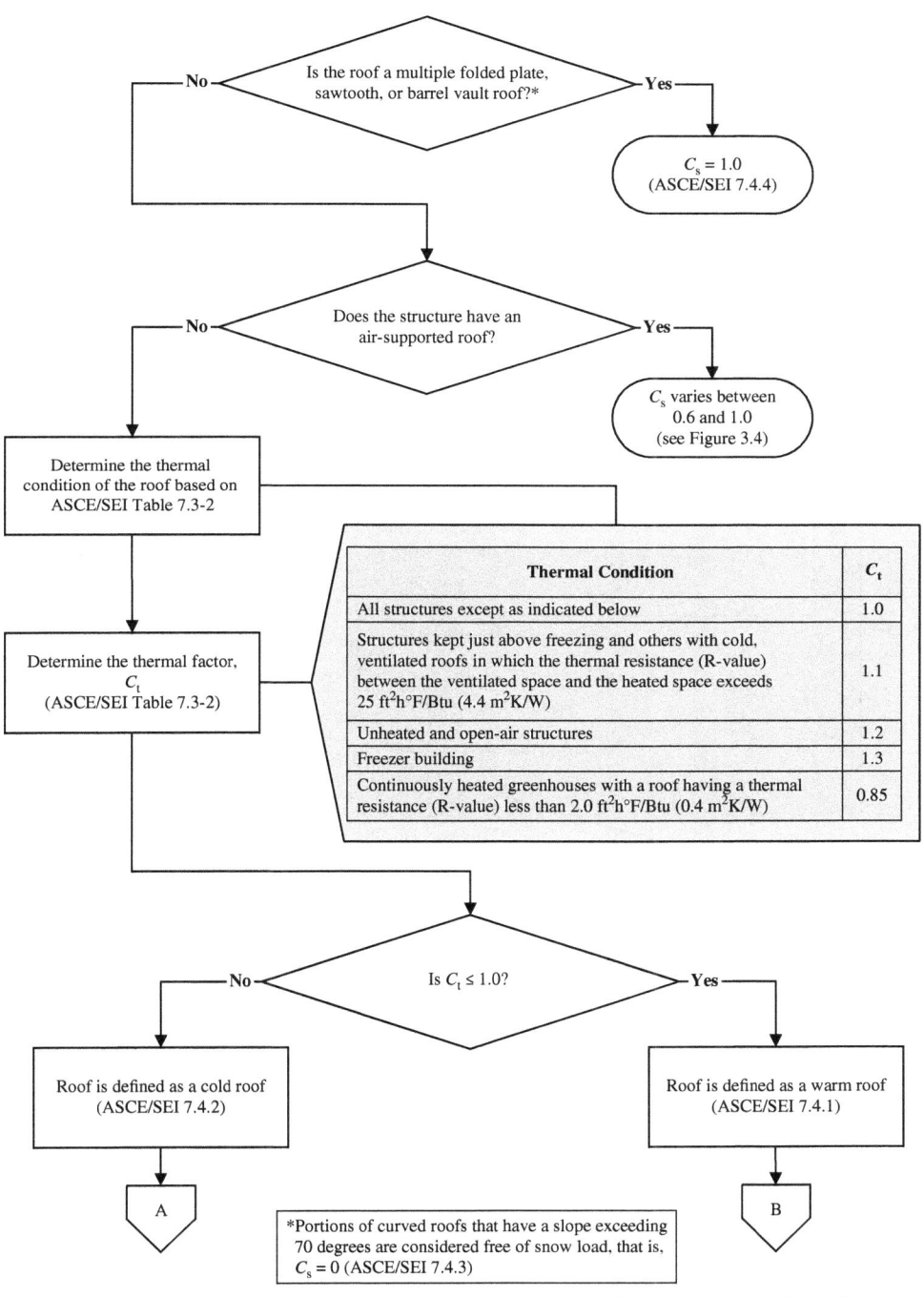

FIGURE 3.5 Flowchart to determine the slope factor, C_s.

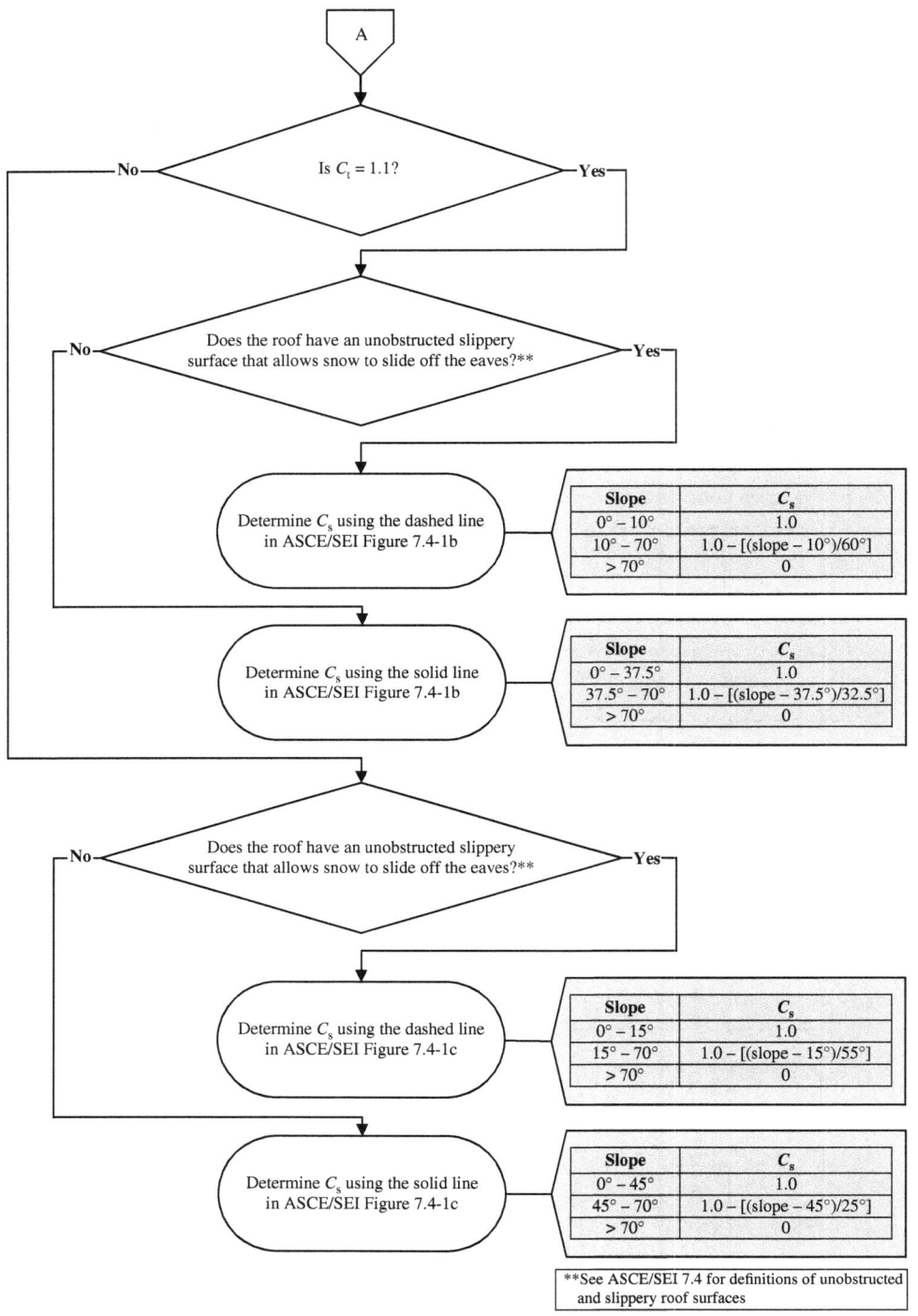

Figure 3.5   *(Continued)*

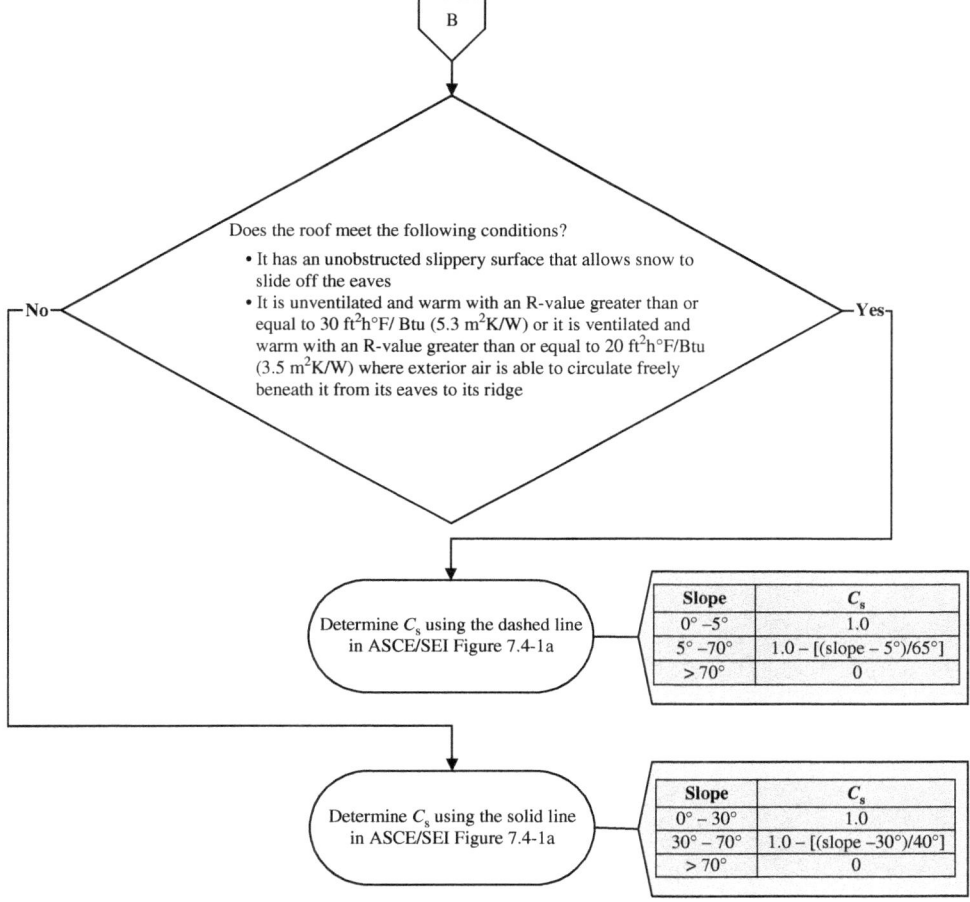

Figure 3.5 *(Continued)*

3.8 Ice Dams and Icicles Along Eaves

Relatively heavy loads due to ice accumulation can occur on the cold, overhanging portions of two types of warm roofs where water drains over their eaves: (1) an unventilated roof with a thermal resistance value (R-value) less than 30 ft²h°F/Btu (5.3 m²K/W) and (2) a ventilated roof with an R-value less than 20 ft²h°F/Btu (3.5 m²K/W).

Illustrated in Fig. 3.6 is the uniformly distributed load, $2p_{f(heated)}$, for ice dams where $p_{f(heated)}$ is the flat roof snow load for the heated portion of the roof, which can be determined using the flowchart in Fig. 3.2. This uniform load is applied over the entire roof overhang where the horizontal extent of the overhang is less than or equal to 5 ft (1.5 m). No other loads except dead loads need to be considered when $2p_{f(heated)}$ is applied to the overhang.

The applicable load case, where the horizontal extent of the overhang is greater than 5 ft (1.5 m), is depicted in Fig. 3.7. The uniform load $2p_{f(heated)}$ need only be applied a distance of 5 ft (1.5 m) from the eave; the remainder of the overhang is subjected to the uniform flat roof snow load for the unheated portion of the roof, $p_{f(unheated)}$.

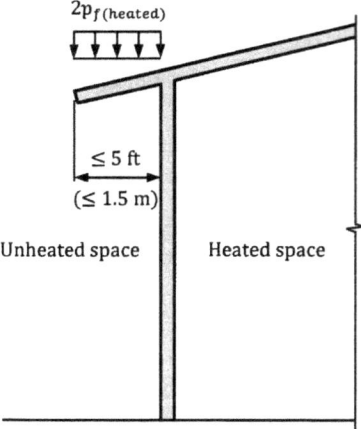

FIGURE 3.6 Load case for ice dams where the horizontal extent of the overhang is less than or equal to 5 ft (1.5 m).

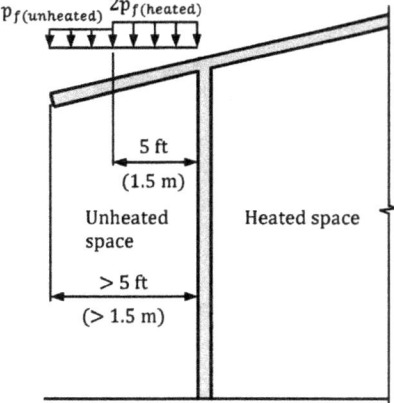

FIGURE 3.7 Load case for ice dams where the horizontal extent of the overhang is greater than 5 ft (1.5 m).

3.9 Partial Loading

For roof framing systems with continuous beams and for other roof systems where removal of snow on one span causes an increase in stress or deflection in an adjacent span, the partial loading provisions of ASCE/SEI 7.5 must be satisfied.

Only the three load cases depicted in Fig. 3.8 need to be investigated for continuous beam systems with and without cantilevered end spans; comprehensive alternate span (checkerboard) loading analyses are not required (see ASCE/SEI Figure 7.5-1):

- Load case 1: Full balanced snow load, p_s, on either of the exterior spans and $p_s/2$ on all other spans.

- Load case 2: $p_s/2$ on either exterior span and p_s on all other spans.

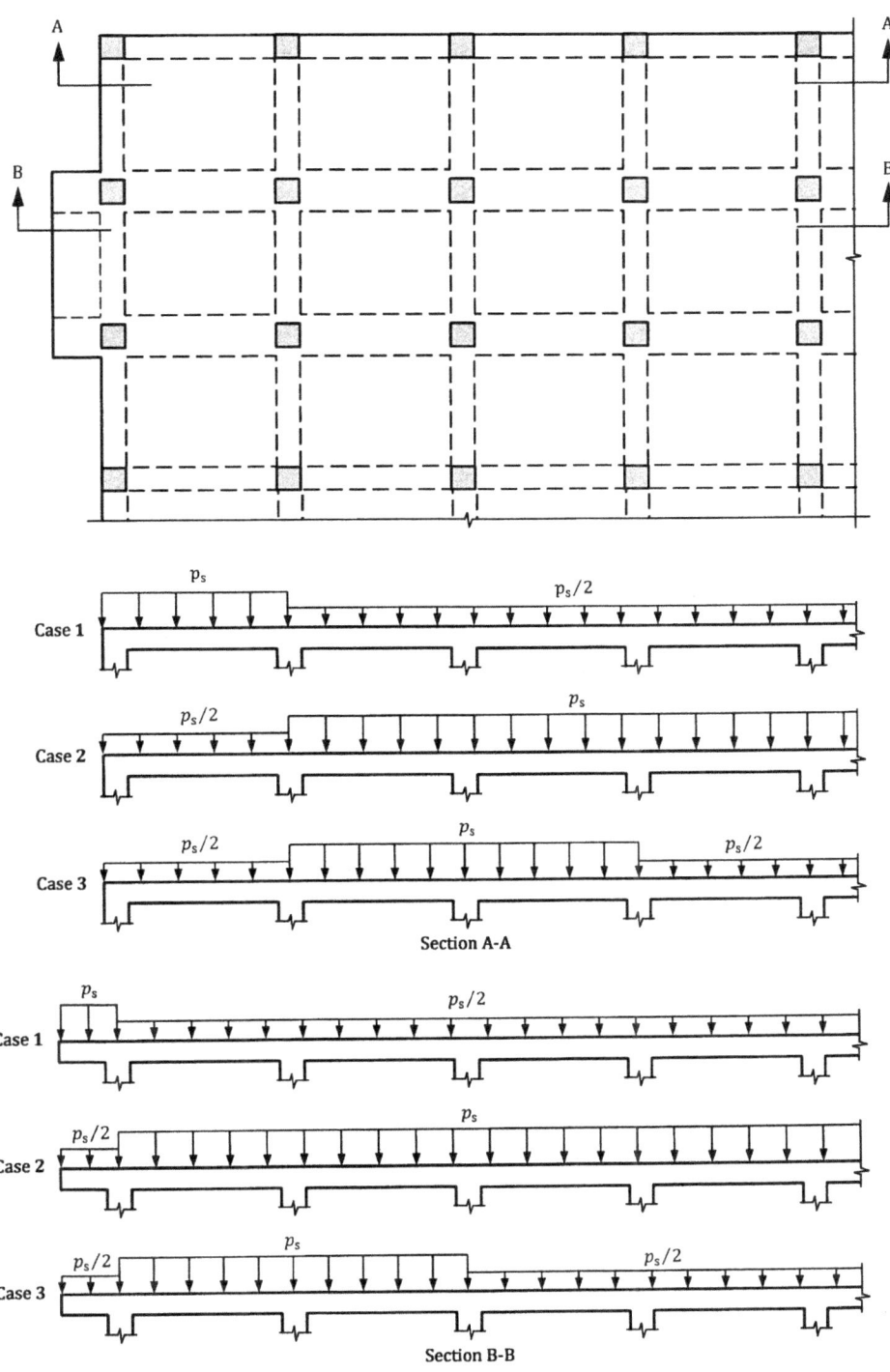

FIGURE 3.8 Partial loading diagrams for continuous beams with or without cantilevered end spans.

- Load case 3: All possible combinations of p_s on any two adjacent spans and $p_s/2$ on all other spans. There are $(n-1)$ possible combinations for this case, where n is equal to the number of spans in the continuous beam system.

Partial loading need not be applied to structural members spanning perpendicular to the ridgeline in gable roofs with slopes between (1) ½ on 12 (2.39 degrees) and (2) 7 on 12 (30.3 degrees).

3.10   Unbalanced Roof Snow Loads

Unbalanced snow loads occur on sloped roofs from wind (which reduces the snow load on the windward portion of the roof and increases it on the leeward portion) and from sunlight (which melts snow on the portions of the roof exposed to sunlight). Unbalanced loads can be considered drift loads. Requirements are given for (1) hip and gable roofs, (2) curved roofs, (3) multiple folded plate, sawtooth, and barrel vault roofs, and (4) dome roofs.

Wind from all directions must be considered when determining unbalanced snow loads. Roofs must be analyzed for balanced and unbalanced loads separately and the structural members are designed for the critical effects from these two cases.

3.10.1   Hip and Gable Roofs

Provisions for unbalanced snow loads on hip and gable roofs are given in ASCE/SEI 7.6.1. Unbalanced snow loads must be considered for roofs with slopes of ½ on 12 (2.39 degrees) through 7 on 12 (30.3 degrees). Snow drifts typically do not form on roofs with slopes less than or greater than these limiting values.

Two unbalanced load conditions are identified in ASCE/SEI 7.6.1. The first of the two conditions is applicable to roofs with an eave to ridge distance, W, less than or equal to 20 ft (6.1 m) where simply supported prismatic members span from the ridge to the eave. The load on the windward portion of the roof is equal to zero and the load on the leeward portion is equal to $I_s p_g$, which is uniformly distributed over the entire width. Where the moment and shear capacities of a structural member (such as a roof truss) varies along its length, it is not prismatic, which means the unbalanced load condition is not applicable in such cases.

The second unbalanced load condition is applicable to all other hip and gable roofs. A uniform load equal to $0.3p_s$ is applied to the windward portion of the roof. The load on the leeward portion consists of two parts: (1) the balanced snow load, p_s, uniformly distributed over the entire width, and (2) a drift load equal to $h_d \gamma / \sqrt{S}$ uniformly distributed over the length $8h_d \sqrt{S}/3$ measured horizontally from the ridge. The drift height, h_d, in this case is determined by the equation in ASCE/SEI Figure 7.6-1 where W is substituted for the length of the roof upwind of the drift, ℓ_u:

$$h_d = \sqrt{I_s} \left\{ [0.43(W)^{1/3}(p_g + 10)^{1/4}] - 1.5 \right\} \quad \text{(ft)} \tag{3.3}$$

$$h_d = \sqrt{I_s} \left\{ [0.42(W)^{1/3}(p_g + 0.479)^{1/4}] - 0.457 \right\} \quad \text{(m)} \tag{3.4}$$

In cases where $W < 20$ ft, use $W = 20$ ft in Eq. (3.3). Similarly, in cases where $W < 6.1$ m, use $W = 6.1$ m in Eq. (3.4). However, h_d need not exceed $\sqrt{I_s p_g W / 4\gamma}$ where W is the actual horizontal distance from the eave to the ridge, not the minimum distance of 20 ft (6.1 m).

The snow density, γ, is determined by ASCE/SEI Equations (7.7-1) and (7.7-1.si):

$$\gamma = 0.13p_g + 14 \le 30 \text{ lb/ft}^3 \tag{3.5}$$

$$\gamma = 0.426p_g + 2.2 \le 4.7 \text{ kN/m}^3 \tag{3.6}$$

The flowchart in Fig. 3.9 can be used to determine balanced and unbalanced snow loads for hip and gable roofs in accordance with ASCE/SEI 7.6.1.

3.10.2 Curved Roofs

Provisions for unbalanced snow loads on curved roofs are given in ASCE/SEI 7.6.2. Any portion of a curved roof where the slope exceeds 70 degrees is considered free of snow loads. The roof slope is measured from the horizontal to the tangent of the curved roof at that point.

Unbalanced snow loads need not be considered where the slope of a straight line from the eaves (or from a point on the roof where the tangent slope is equal to 70 degrees) to the crown is less than 10 degrees or is greater than 60 degrees.

The provisions in ASCE/SEI 7.6.2 are applicable to curved roofs that are concave downward. For other roof geometries, such as a concave upward curved roof, or for complicated site conditions, wind tunnel models should be used to establish design snow loads (see ASCE/SEI C7.14).

Three balanced and unbalanced load cases are identified in ASCE/SEI Figure 7.4-2:

Case 1: Slope at eaves less than 30 degrees
In this case, the balanced load is trapezoidal near the eaves and uniform over a segment centered on the crown. The magnitude of the balanced load, p_s, is equal to $C_s p_f$ where C_s is determined from Fig. 3.5 and p_f is determined from Fig. 3.2. At the eaves, $C_{s|eave}$ is determined using the slope of the roof at those locations. For shallow curved roofs, C_s may be equal to 1.0 over the entire roof.

The windward portion of the roof between the windward eave and the crown is assumed to be free of snow in the unbalanced load case. A trapezoidal load varying from $0.5p_f$ at the crown to $2p_f(C_{s|eave}/C_e)$ at the eave occurs on the leeward portion of the roof where $C_{s|eave}$ is determined using the slope of the roof at the eave and the exposure factor, C_e, is determined from Fig. 3.2.

Case 2: Slope at eaves 30 degrees to 70 degrees
The balanced load in this case is equal to p_f over the segment of the roof centered at the crown with two sets of trapezoidal loads on each side of this segment. Where the roof slope is equal to 30 degrees, the magnitude of the load is $C_{s|30}p_f$ with the roof slope factor $C_{s|30}$ determined from Fig. 3.5. Similarly, the load at the eaves is equal to $C_{s|eave}p_f$ where $C_{s|eave}$ is determined from Fig. 3.5 based on the roof angle at that location.

Like in case 1, the windward portion of the roof between the windward eave and the crown is assumed to be free of snow in the unbalanced load case. Two cases must be considered for the leeward portion. Where the ground or another roof abuts the curved roof at or within 3 ft (0.9 m) of its eaves, the unbalanced load is $0.5p_f$ at the crown and $2p_f(C_{s|30}/C_e)$ at the point where the roof slope is equal to 30 degrees. From that point to the eaves, the load is a constant $2p_f(C_{s|30}/C_e)$. In all other cases, the unbalanced loading is the same as that described above except for the segment of the roof between the point where the slope is 30 degrees and the eaves; in that segment, the load at the 30-degree point is equal to $2p_f(C_{s|30}/C_e)$, which decreases linearly to $2p_f(C_{s|eave}/C_e)$.

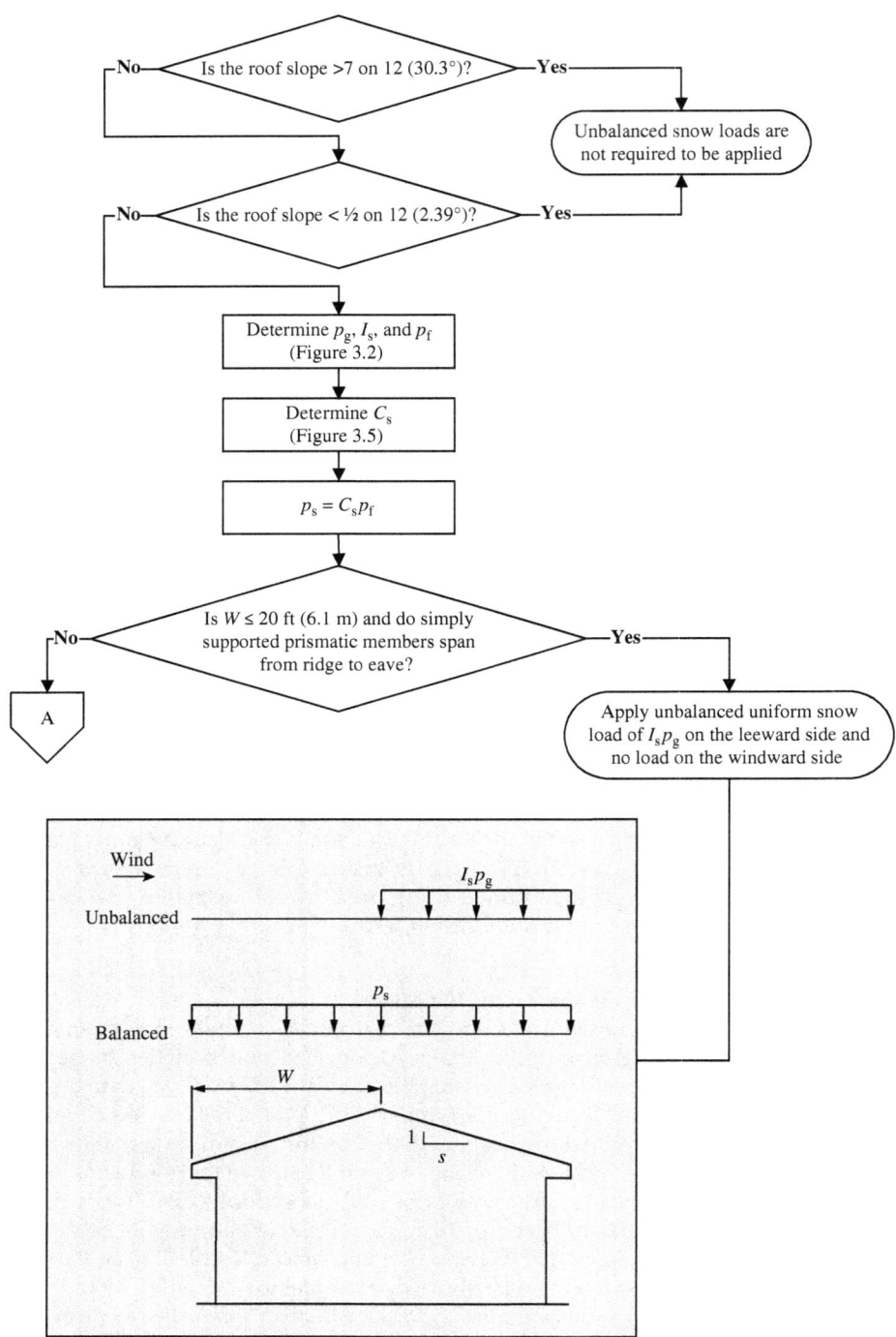

FIGURE 3.9 Flowchart to determine balanced and unbalanced snow loads on hip and gable roofs.

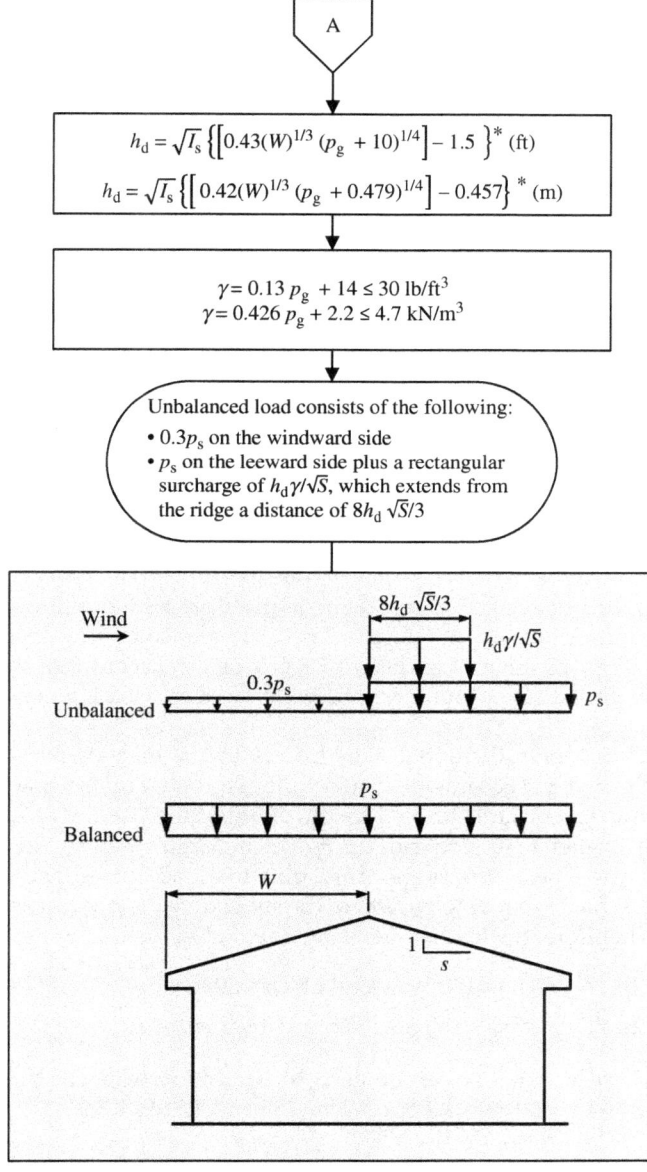

$$h_d = \sqrt{I_s}\left\{\left[0.43(W)^{1/3}\,(p_g+10)^{1/4}\right]-1.5\right\}^* \text{ (ft)}$$

$$h_d = \sqrt{I_s}\left\{\left[0.42(W)^{1/3}\,(p_g+0.479)^{1/4}\right]-0.457\right\}^* \text{ (m)}$$

$$\gamma = 0.13\,p_g + 14 \le 30 \text{ lb/ft}^3$$
$$\gamma = 0.426\,p_g + 2.2 \le 4.7 \text{ kN/m}^3$$

Unbalanced load consists of the following:
- $0.3p_s$ on the windward side
- p_s on the leeward side plus a rectangular surcharge of $h_d\gamma/\sqrt{S}$, which extends from the ridge a distance of $8h_d\sqrt{S}/3$

* Where $W < 20$ ft (6.1 m), use $W = 20$ ft (6.1 m) in this equation. However, h_d need not exceed $\sqrt{I_s p_g\, W/4\gamma}$ where W is the actual horizontal distance from the eave to the ridge, not the minimum distance of 20 ft (6.1 m).

FIGURE 3.9 (Continued)

Case 3: Slope at eaves greater than 70 degrees

The balanced load is equal to p_f over the segment of the roof centered at the crown with trapezoidal and triangular loads on each side of this segment. The trapezoidal loads extend from the points where $C_s = 1.0$ to the points where the roof slope is 30 degrees. At the 30-degree points, the load is $C_{s|30}p_f$. The triangular loads extend from the points where the roof slope is 30 degrees to the points where it is 70 degrees. The load is zero in the segments between the points where the roof slope is 70 degrees and the eaves.

The unbalanced load for this case is very similar to that for case 2: The windward portion of the roof between the windward eave and the crown is assumed to be free of snow, and there are two cases to consider for the leeward portion depending on whether the ground or another roof abuts the curved roof within 3 ft (0.9 m) of its eaves.

The flowchart in Fig. 3.10 can be used to determine balanced and unbalanced snow loads for curved roofs in accordance with ASCE/SEI 7.6.2.

3.10.3   Multiple Folded Plate, Sawtooth, and Barrel Vault Roofs

Provisions for unbalanced snow loads on multiple folded plate, sawtooth, and barrel vault roofs are given in ASCE/SEI 7.6.3. Unbalanced snow loads must be applied where the slope of the roof exceeds 1.79 degrees. In accordance with ASCE/SEI 7.4.4, $C_s = 1.0$ for these types of roofs, so the balanced snow load, p_s, is equal to the flat roof snow load, p_f.

Like curved roofs, the unbalanced load is $0.5p_f$ at the crown or ridge of the roof and $2p_f/C_e$ at the valley. The snow load at a valley is limited by the space available for snow accumulation (that is, it is limited by the depth of the valley). Assume the vertical distance from a valley to the roof ridge is h_r. The maximum snow load occurring at a valley is equal to the load at the ridge, $0.5p_f$, plus the load corresponding to a snow depth equal to h_r, which is equal to γh_r where the snow density, γ, is determined by Eq. (3.5) or Eq. (3.6). Therefore, the load at the valley is equal to the following:

- If $2p_f/C_e < 0.5p_f + \gamma h_r$, the load at the valley $= 2p_f/C_e$
- If $2p_f/C_e \geq 0.5p_f + \gamma h_r$, the load at the valley $= 0.5p_f + \gamma h_r$

The flowchart in Fig. 3.11 can be used to determine balanced and unbalanced snow loads for a sawtooth roof in accordance with ASCE/SEI 7.6.3.

3.10.4   Dome Roofs

According to ASCE/SEI 7.6.4, unbalanced snow loads on dome roofs are determined in the same manner as for curved roofs. Unbalanced loads are to be applied to the downwind 90-degree sector of the dome in plan (see Fig. 3.12). The load decreases linearly to zero over 22.5-degree sectors on each side of the 90-degree sector. No snow load is taken on the remaining 225-degree upwind sector.

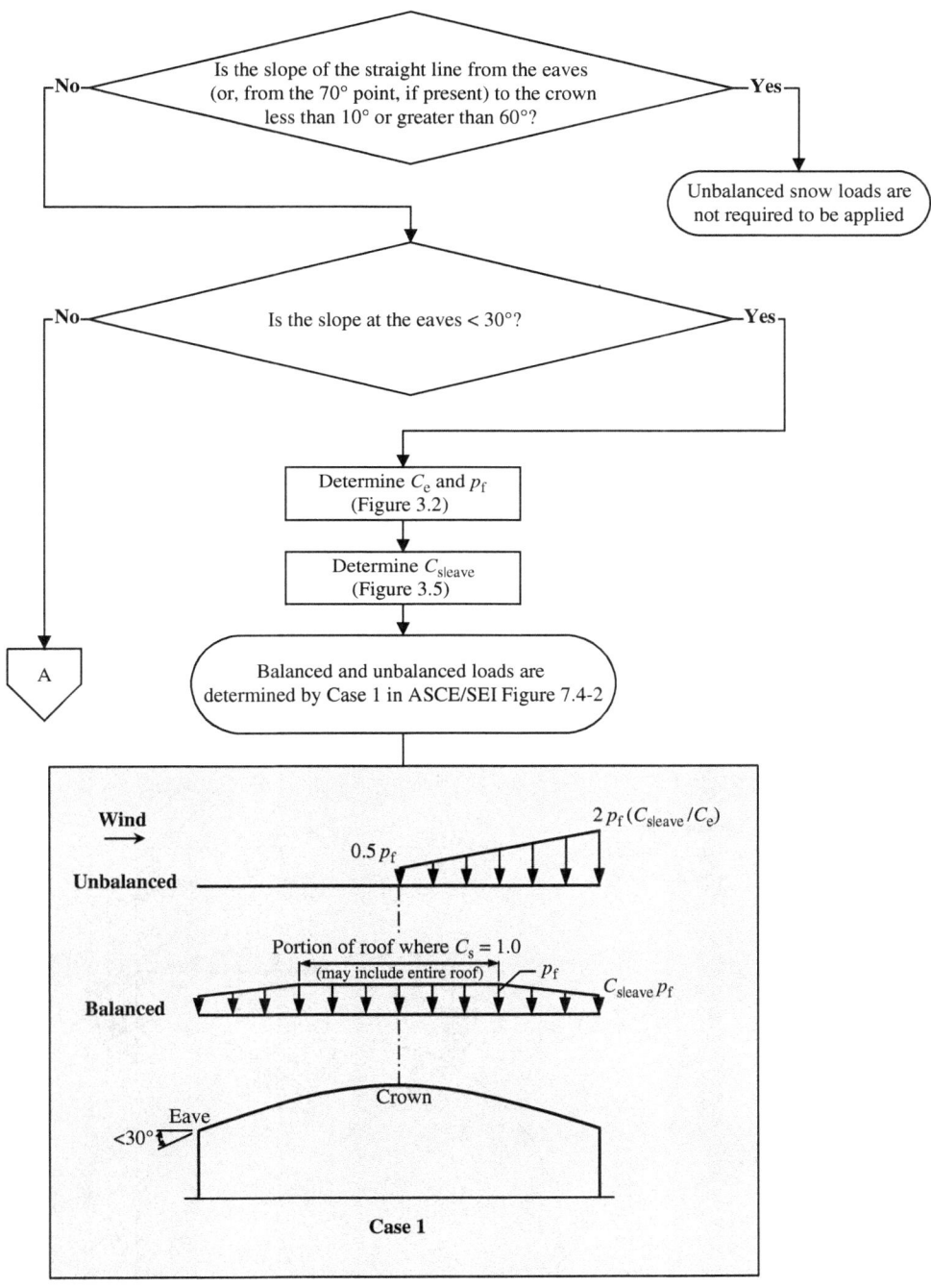

FIGURE 3.10 Flowchart to determine balanced and unbalanced snow loads on curved roofs.

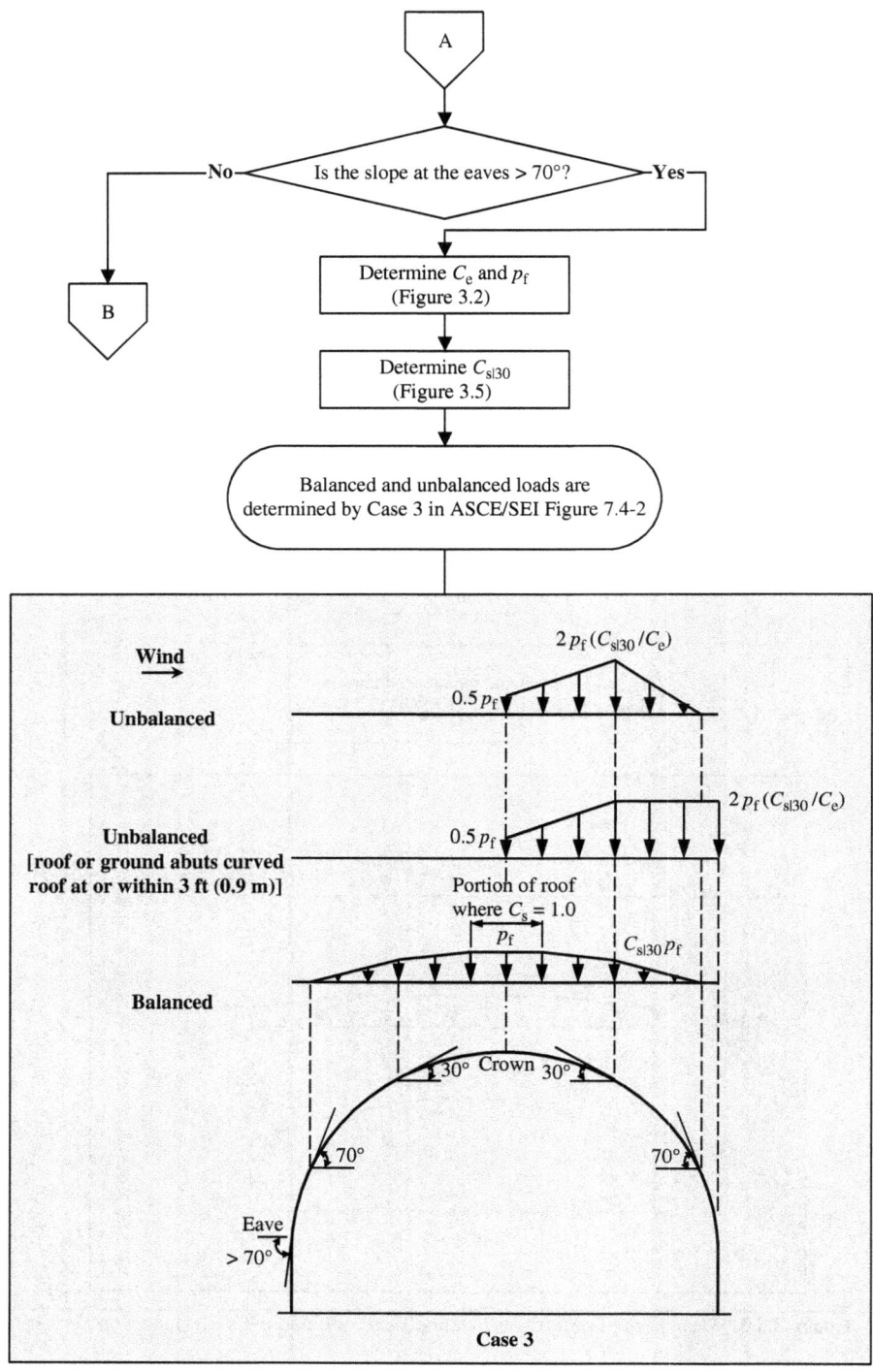

FIGURE 3.10 *(Continued)*

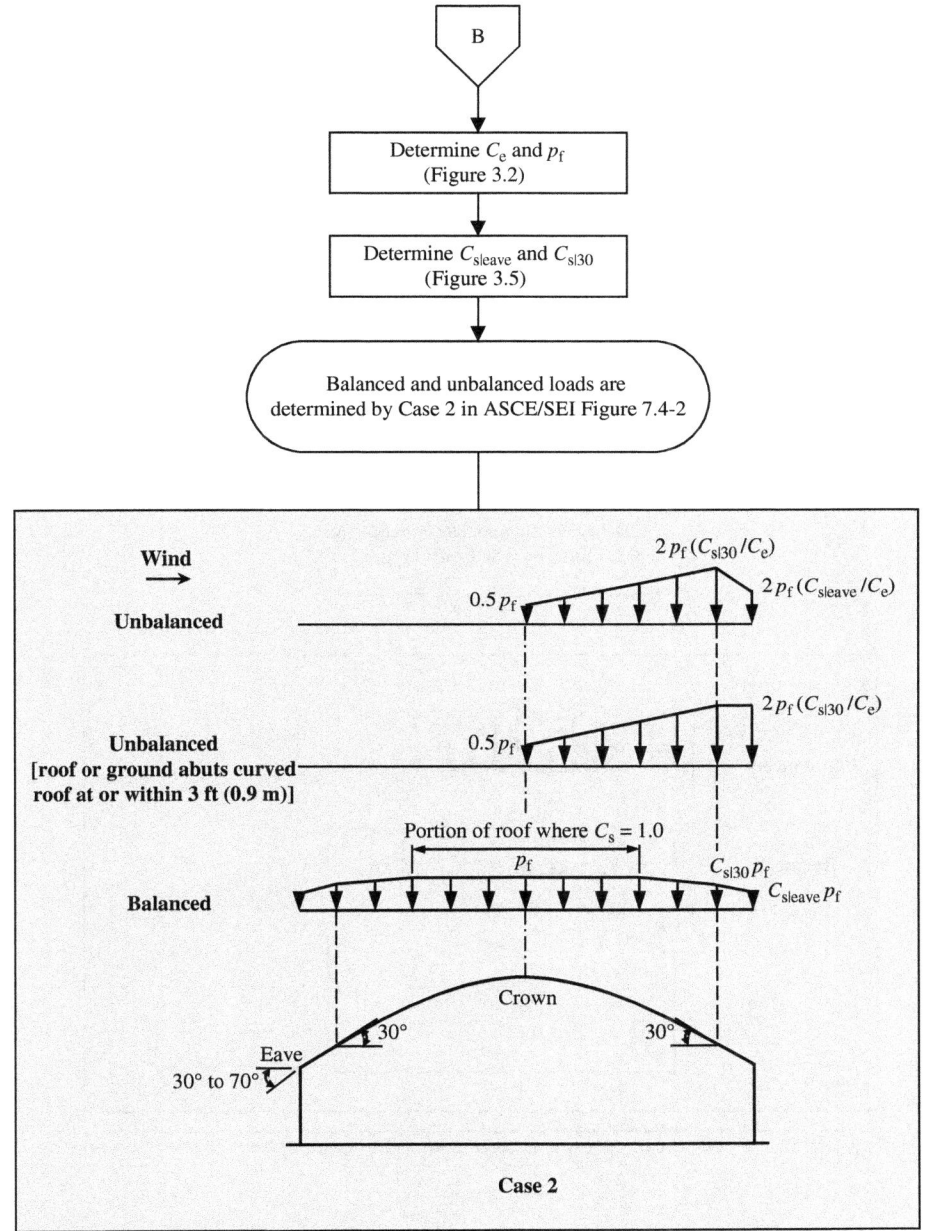

FIGURE 3.10 (*Continued*)

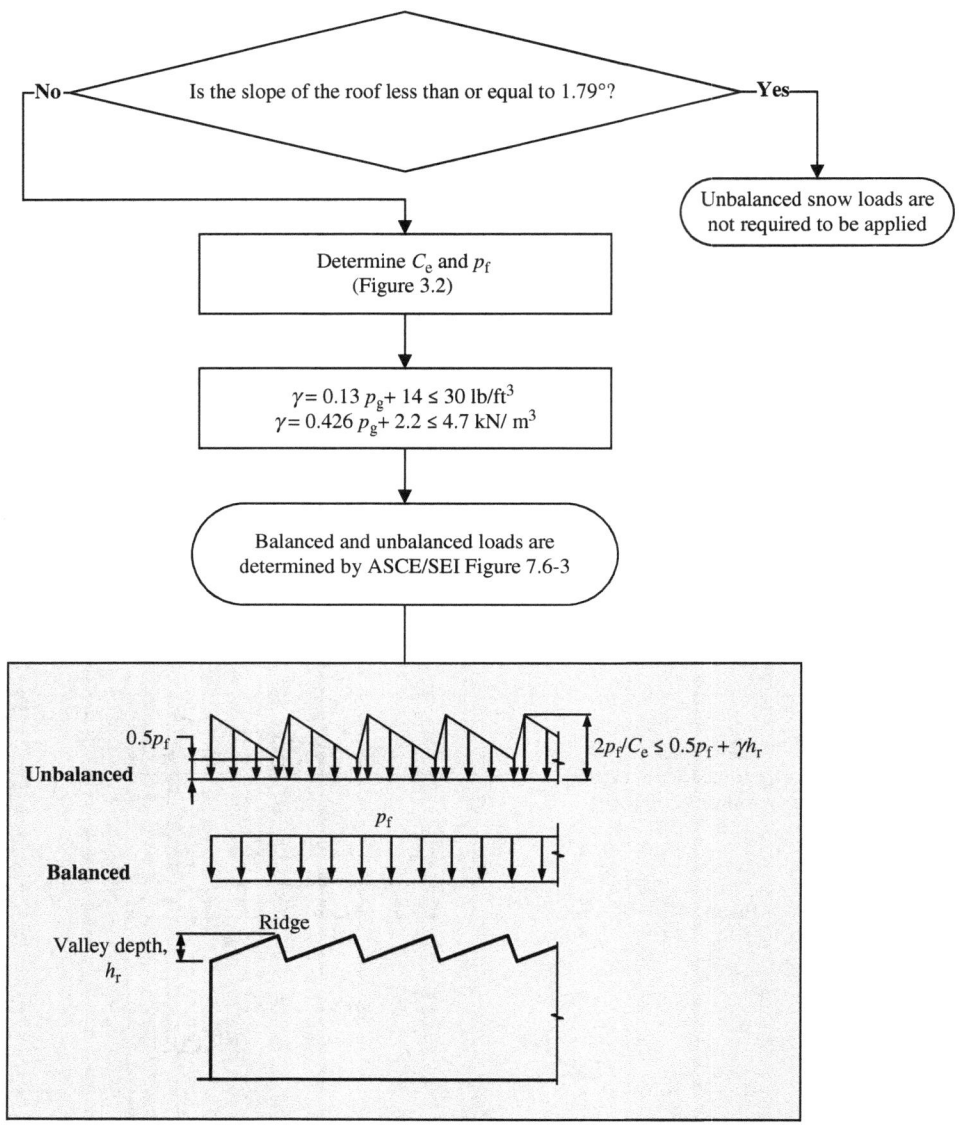

Figure 3.11 Flowchart to determine balanced and unbalanced snow loads on sawtooth roofs.

The balanced and unbalanced load distributions for a dome roof can be determined using Fig. 3.10. In the unbalanced load case, the snow load at the eave or at the location where the roof slope is equal to 70 degrees decreases linearly to zero over the 22.5-degree sector on each side of the 90-degree downwind sector of the roof.

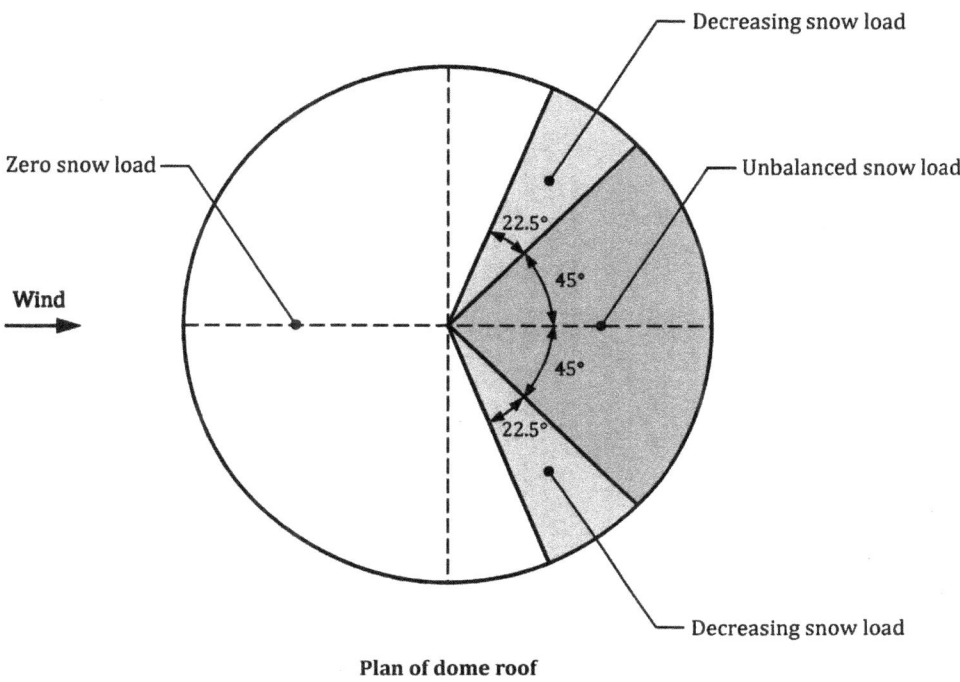

Plan of dome roof

FIGURE 3.12 Unbalanced snow loads for a dome roof.

3.11 Drifts on Lower Roofs

Provisions for snow drifts occurring on lower roofs of a building are given in ASCE/SEI 7.7. Two types of drifts are addressed:

- Leeward drifts: Wind deposits snow from (a) a higher portion of the same building or an adjacent building or (b) a terrain feature (such as a hill) to a lower roof.

- Windward drifts: Wind deposits snow from the windward portion of a lower roof to the portion of the lower roof adjacent to a taller part of the building.

Both types of drifts are covered in the following sections. Also covered are provisions for drifting on lower structures adjacent to taller ones and intersecting drifts at lower roofs.

3.11.1 Lower Roof a Structure

Depending on wind direction, either a windward or leeward drift can form on a lower roof. The configuration for snow drifts on lower roofs is given in ASCE/SEI 7.7.1 (see Fig. 3.13).

Drift loads need not be considered where the following equation is satisfied:

$$h_c/h_b = (h_{step} - h_b)/h_b = [h_{step}/(p_s/\gamma)] - 1 < 0.2 \qquad (3.7)$$

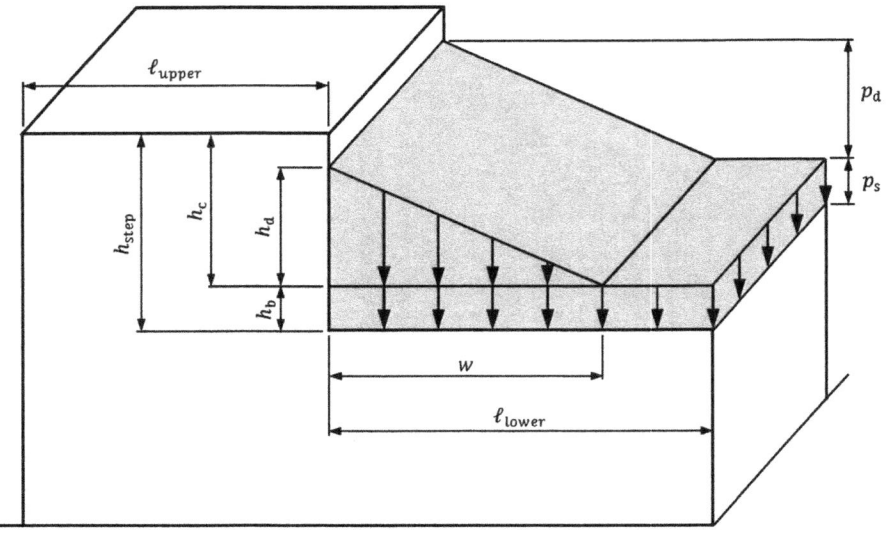

FIGURE 3.13   Drift configuration on a lower roof.

In this equation, h_c is the clear height from the top of the balanced snow load to the closet point on the adjacent upper roof (see Fig. 3.13). This length can be obtained by subtracting the height of the balanced snow load, $h_b = p_s/\gamma$, from h_{step}, which is the vertical distance from the lower roof to the upper roof.

Where drift loads must be considered, the total snow load at the step, p_{total}, is equal to the drift load, p_d, which is assumed to be triangular, plus the uniform balanced snow load, p_s. The maximum drift load is determined by the following equation:

$$p_d = \gamma h_d \quad (\text{lb} / \text{ft}^2, \text{kN/m}^2) \tag{3.8}$$

where the snow density, γ, is determined by Eqs. (3.5) and (3.6). The drift height, h_d, to use in Eq. (3.8) is the larger of the leeward and windward drift heights, $h_{d,\text{leeward}}$ and $h_{d,\text{windward}}$, determined in accordance with ASCE/SEI Figure 7.6-1 (ASCE/SEI 7.7.1). Equations to determine $h_{d,\text{leeward}}$ and $h_{d,\text{windward}}$ based on these requirements are given in Table 3.5.

The maximum drift load, p_d, the total load, p_{total}, and the width of the drift on the lower roof, w, depends on the clear height, h_c, and the calculated value of h_d:

- Where $h_d \le h_c = h_{step} - h_b$:

$$p_d = \gamma h_d \text{ and } w = 4h_d$$

$$p_{total} = p_s + p_d = \gamma(h_b + h_d)$$

- Where $h_d > h_c = h_{step} - h_b$:

$$p_d = \gamma h_c \text{ and } w = 4h_d^2/h_c \le 8h_c$$

$$p_{total} = p_s + p_d = \gamma(h_b + h_c) = \gamma h_{step}$$

	ℓ_{upper}	$h_{d,leeward}$
Leeward drift	≥ 20 ft	$\sqrt{I_s}\left\{[0.43(\ell_{upper})^{1/3}(p_g+10)^{1/4}]-1.5\right\}\leq 0.6\ell_{lower}$
	≥ 6.1 m	$\sqrt{I_s}\left\{[0.42(\ell_{upper})^{1/3}(p_g+0.479)^{1/4}]-0.457\right\}\leq 0.6\ell_{lower}$
	< 20 ft	$\sqrt{I_s}\left\{[0.43(20)^{1/3}(p_g+10)^{1/4}]-1.5\right\}\leq \sqrt{I_sp_g\ell_{upper}/4\gamma}$
	< 6.1 m	$\sqrt{I_s}\left\{[0.42(6.1)^{1/3}(p_g+0.479)^{1/4}]-0.457\right\}\leq \sqrt{I_sp_g\ell_{upper}/4\gamma}$
	ℓ_{lower}	$h_{d,windward}$
Windward drift	≥ 20 ft	$0.75\sqrt{I_s}\left\{[0.43(\ell_{lower})^{1/3}(p_g+10)^{1/4}]-1.5\right\}$
	≥ 6.1 m	$0.75\sqrt{I_s}\left\{[0.42(\ell_{lower})^{1/3}(p_g+0.479)^{1/4}]-0.457\right\}$
	< 20 ft	$0.75\sqrt{I_s}\left\{[0.43(20)^{1/3}(p_g+10)^{1/4}]-1.5\right\}\leq \sqrt{I_sp_g\ell_{lower}/4\gamma}$
	< 6.1 m	$0.75\sqrt{I_s}\left\{[0.42(6.1)^{1/3}(p_g+0.479)^{1/4}]-0.457\right\}\leq \sqrt{I_sp_g\ell_{lower}/4\gamma}$

TABLE 3.5 Equations for Determining Drift Height

In cases where $w > \ell_{lower}$, the drift load is to taper linearly to zero at the far end of the lower roof (see Fig. 3.14). This provision may be applicable to canopies over an entrance-way to a building, for example.

The flowchart in Fig. 3.15 can be used to determine drift loads and widths for lower roofs in accordance with ASCE/SEI 7.7.1.

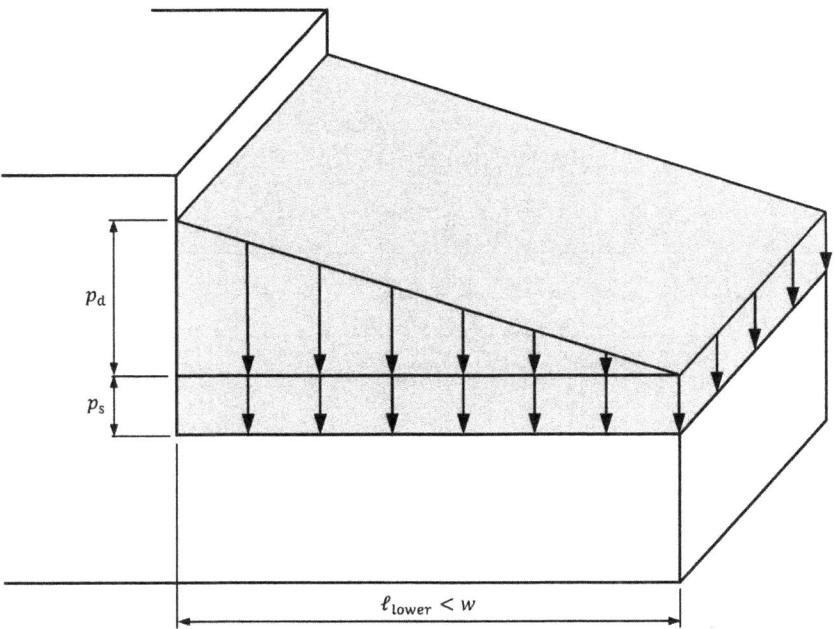

FIGURE 3.14 Load configuration where the drift width is greater than the length of the lower roof.

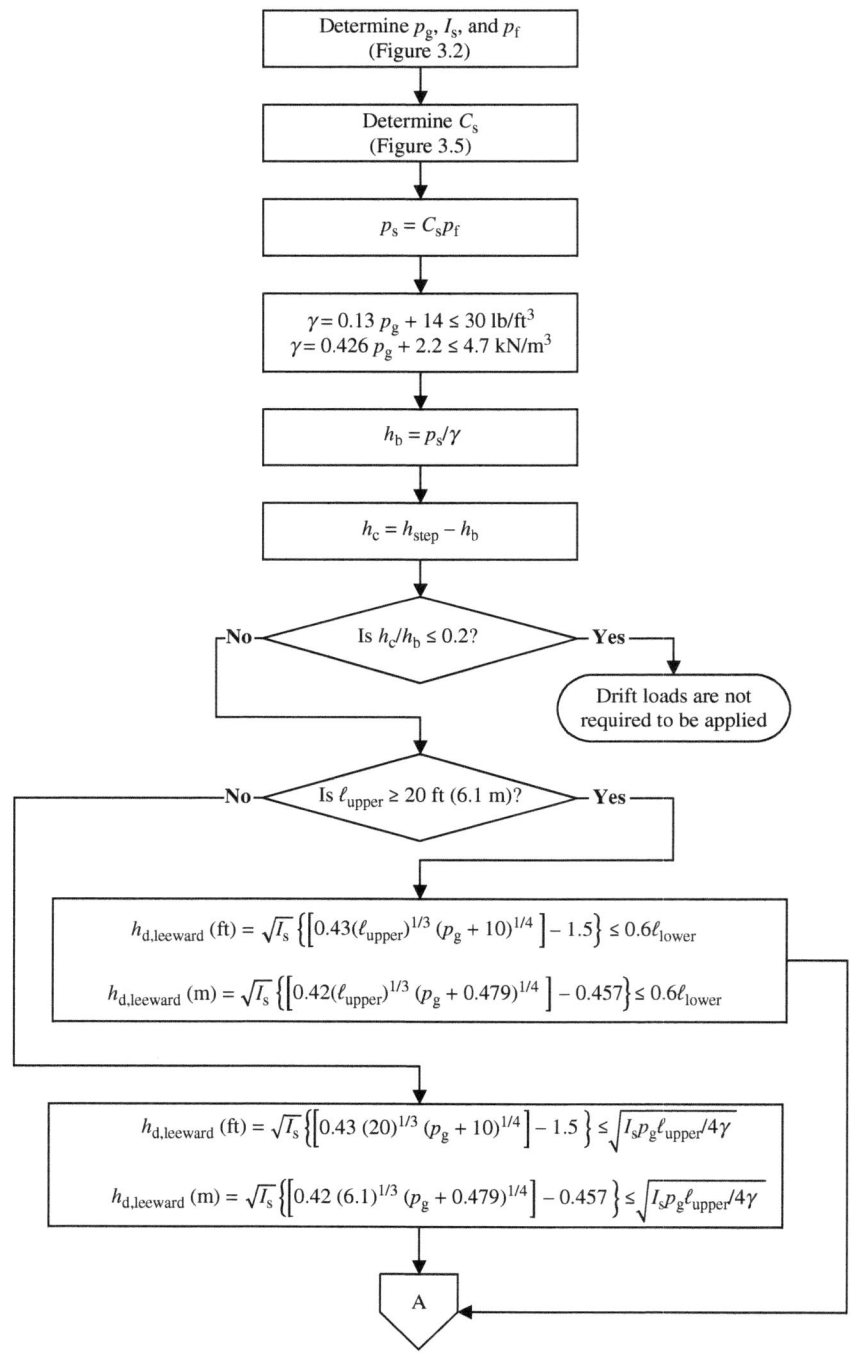

FIGURE 3.15 Flowchart to determine drift loads and widths on a lower roof of a structure.

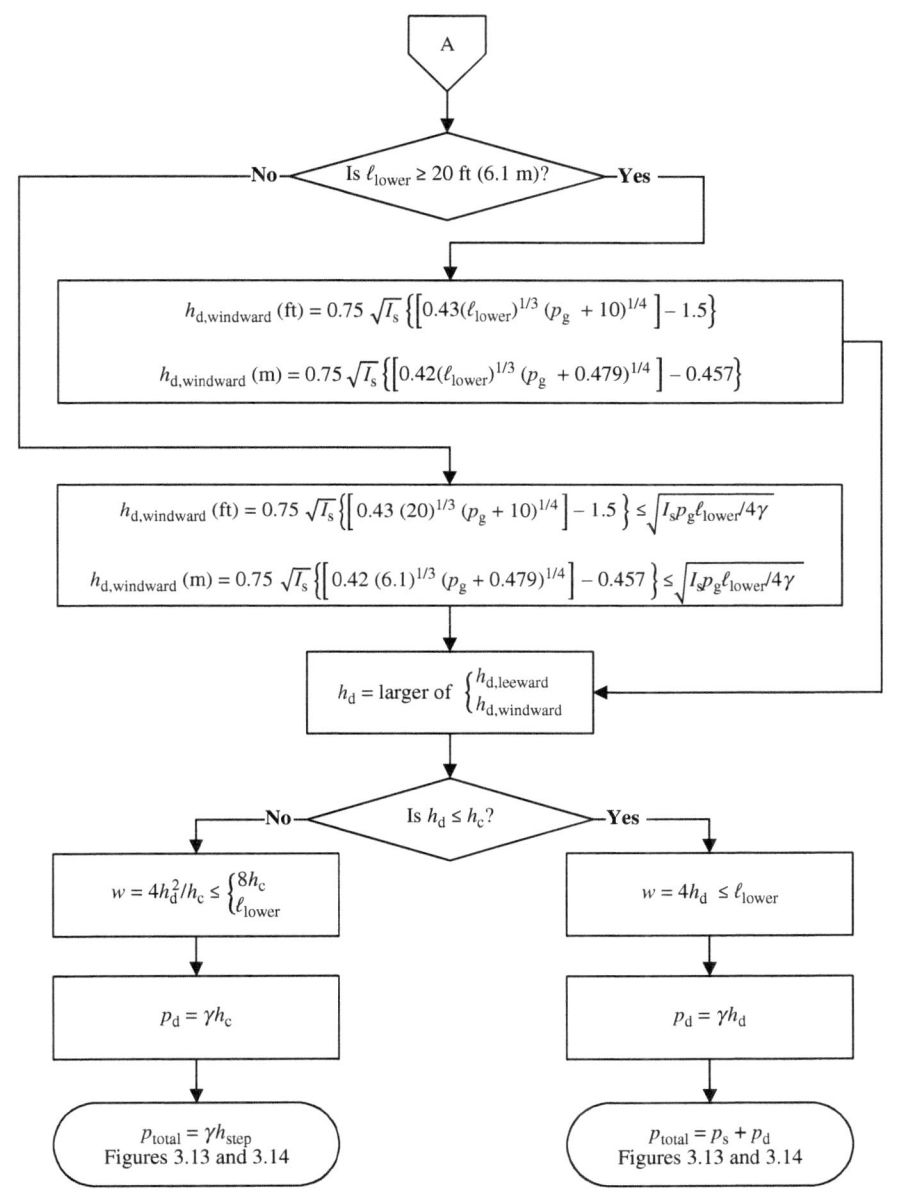

FIGURE 3.15   (Continued)

3.11.2   Adjacent Structures

Leeward and windward drifts can form on the roof of a structure adjacent to a structure with a higher roof. Leeward drifts can form on lower roofs when the horizontal separation distance, s, between the adjacent structures is less than 20 ft (6.1 m) and less than $6h$ where h is the vertical separation distance between the two structures (ASCE/SEI 7.7.2). The drift load, p_d, in such cases is determined by the provisions of ASCE/SEI 7.7.1 using a drift height equal to the smaller of the following: (1) $h_{d,\text{leeward}}$ based on ℓ_{upper} and (2) $(6h - s)/6$. The drift width, w, on the lower roof is equal to 6 times the smaller of the drift heights, which are equal to $6h_{d,\text{leeward}}$ and $(6h - s)$.

Windward drifts on the lower roof are also determined by the provisions of ASCE/SEI 7.7.1. It is assumed $h_{d,\text{windward}}$ occurs at the face of the adjacent higher structure and is calculated using ℓ_{lower}. Within the separation zone between the two structures, the drift is truncated, and the height of the drift at the edge of the lower structure can be determined from geometry. The total drift width in this case is equal to $4h_{d,\text{windward}}$, which extends from the face of the structure with the higher roof to the point on the lower roof where $p_d = 0$.

The flowchart in Fig. 3.16 can be used to determine leeward and windward drifts on an adjacent lower roof in accordance with ASCE/SEI 7.7.2.

3.11.3   Intersecting Drifts on Low Roofs

Intersecting drifts can form at locations such as reentrant corners and parapet wall corners due to wind acting in multiple directions. Provisions for intersecting snow drifts are given in ASCE/SEI 7.7.3. The provisions in ASCE/SEI 7.7.1 are used to determine the individual drift geometries for both leeward and windward intersecting drifts.

Intersecting drift loads are considered to occur concurrently and are combined as shown in Fig. 3.17. The drift load at the intersection is based on the larger drift height and not on the addition of the two drift heights. Also shown in Fig. 3.17 are equations to determine the leeward and windward drift heights based on the provisions in ASCE/SEI 7.7.1 assuming all the upper and lower lengths are greater than or equal to 20 ft (6.1 m). These drift heights and the corresponding drift loads and widths can be determined using the flowchart in Fig. 3.15.

3.12   Roof Projections and Parapets

Drift loads on roof projections (including rooftop equipment) and parapet walls are determined by the provisions in ASCE/SEI 7.8 and are based on ASCE/SEI 7.7.1. Windward drifts are formed on the side of parapet walls. Both windward and leeward drifts can form adjacent to rooftop units and other projections; however, the leeward drift is typically insignificant so for simplicity, only windward drifts are considered.

For roof projections, the height of the drift is equal to three-quarters of the value of h_d determined by the equation in ASCE/SEI Figure 7.6-1 where ℓ_u is equal to the larger of the length of the roof upwind or downwind of the projection in that direction. Drift loads are not required where the side of a projection is less than 15 ft (4.6 m) long or where the clear distance between the height of the balanced snow load, $h_b = p_s/\gamma$, and the bottom of the projection (including horizontal supports) is greater than or equal to 2 ft (0.61 m).

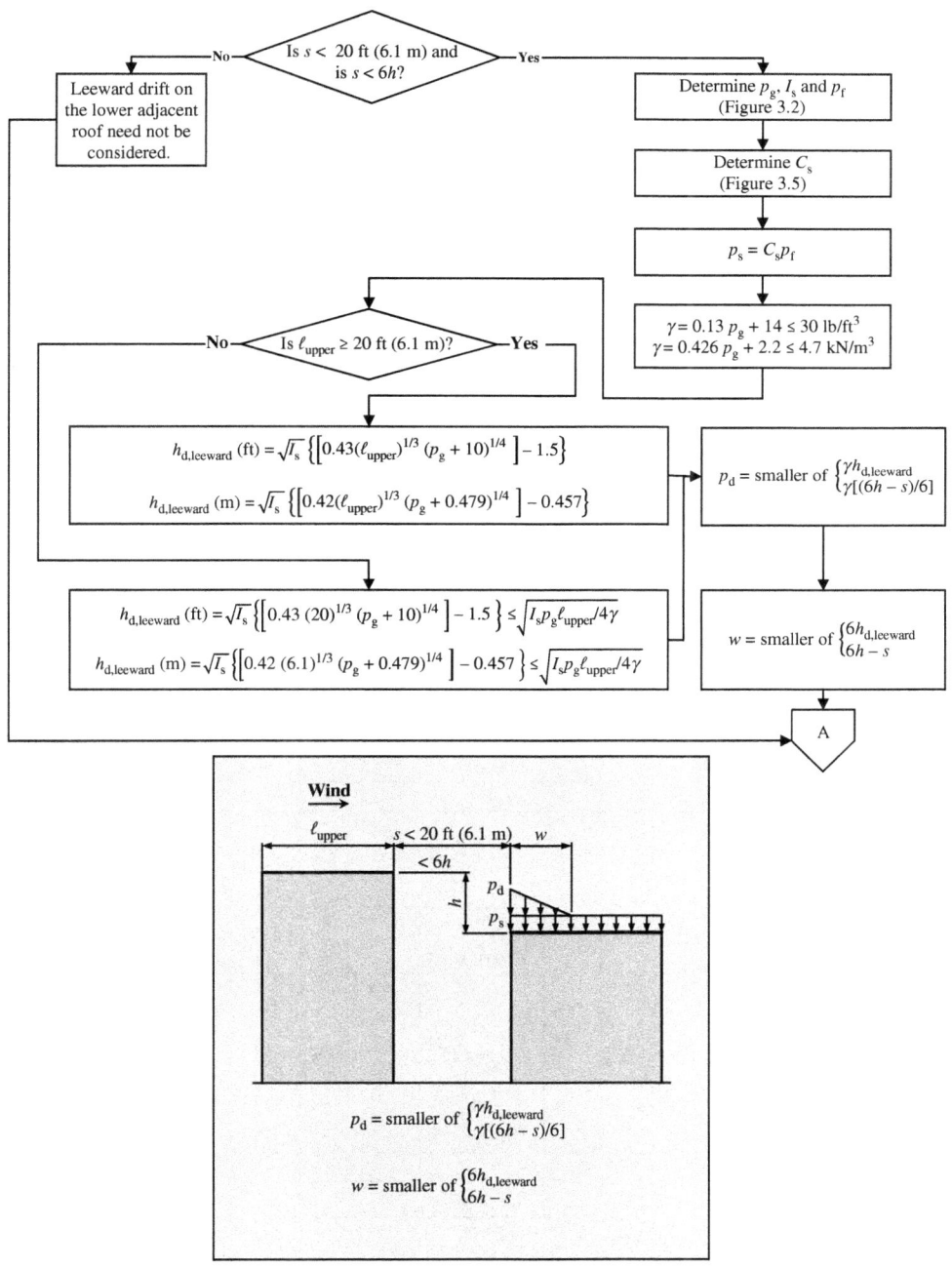

FIGURE 3.16 Flowchart to determine leeward and windward drifts on an adjacent lower roof.

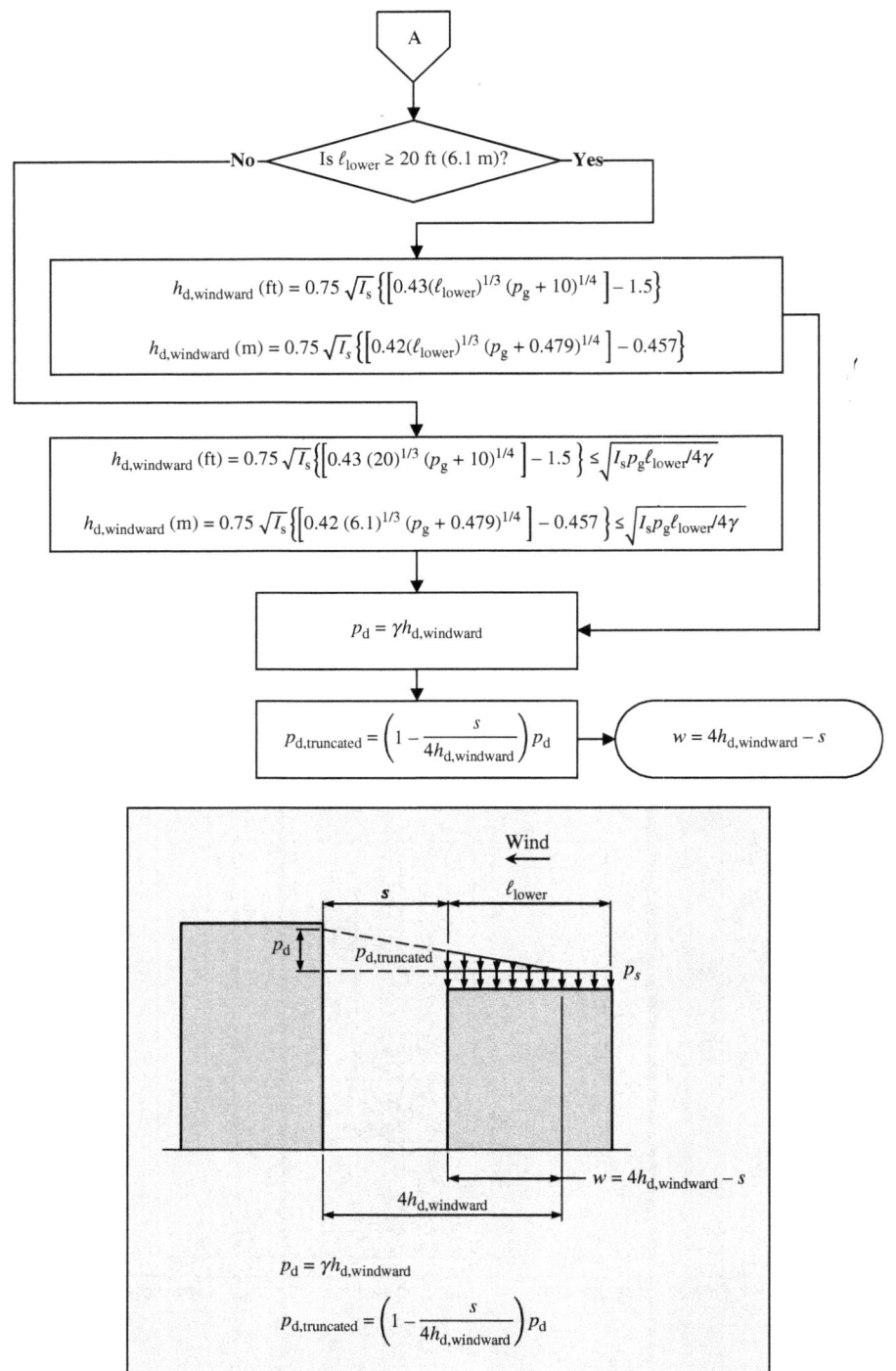

A

Is $\ell_{lower} \geq 20$ ft (6.1 m)? No / Yes

$$h_{d,windward} \text{ (ft)} = 0.75 \sqrt{I_s} \left\{ \left[0.43(\ell_{lower})^{1/3} (p_g + 10)^{1/4} \right] - 1.5 \right\}$$

$$h_{d,windward} \text{ (m)} = 0.75 \sqrt{I_s} \left\{ \left[0.42(\ell_{lower})^{1/3} (p_g + 0.479)^{1/4} \right] - 0.457 \right\}$$

$$h_{d,windward} \text{ (ft)} = 0.75 \sqrt{I_s} \left\{ \left[0.43\,(20)^{1/3} (p_g + 10)^{1/4} \right] - 1.5 \right\} \leq \sqrt{I_s p_g \ell_{lower}/4\gamma}$$

$$h_{d,windward} \text{ (m)} = 0.75 \sqrt{I_s} \left\{ \left[0.42\,(6.1)^{1/3} (p_g + 0.479)^{1/4} \right] - 0.457 \right\} \leq \sqrt{I_s p_g \ell_{lower}/4\gamma}$$

$$p_d = \gamma h_{d,windward}$$

$$p_{d,truncated} = \left(1 - \frac{s}{4h_{d,windward}} \right) p_d$$

$$w = 4h_{d,windward} - s$$

Wind

s ℓ_{lower}

p_d

$p_{d,truncated}$ p_s

$w = 4h_{d,windward} - s$

$4h_{d,windward}$

$$p_d = \gamma h_{d,windward}$$

$$p_{d,truncated} = \left(1 - \frac{s}{4h_{d,windward}} \right) p_d$$

FIGURE 3.16 *(Continued)*

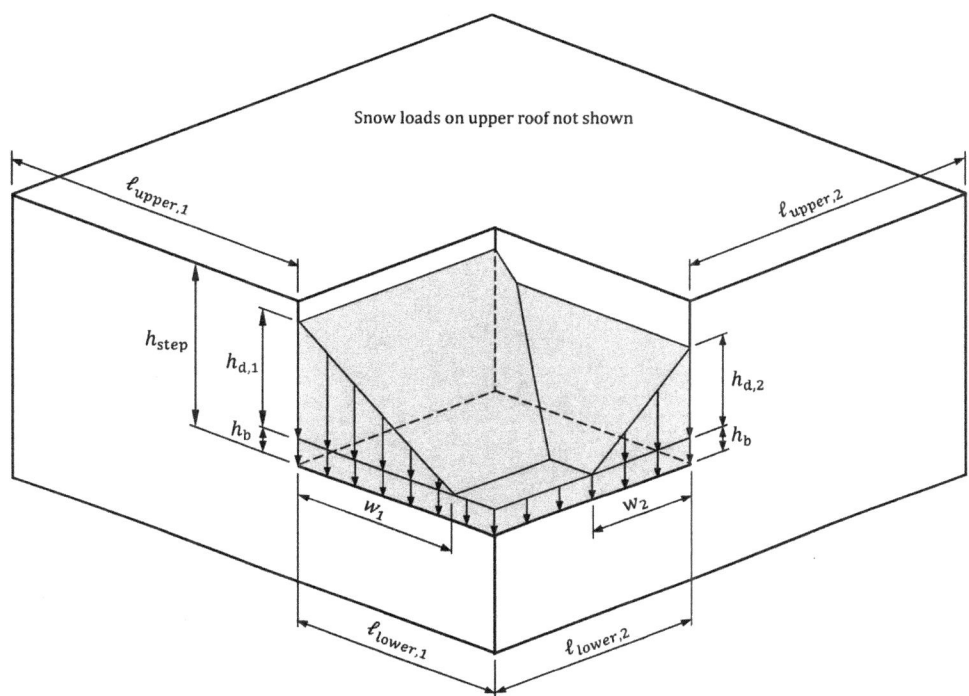

Snow loads on upper roof not shown

Leeward drifts
$h_{d,1}$ (ft) $= \sqrt{I_s} \left\{ \left[0.43 \left(\ell_{upper,1} \right)^{1/3} \left(p_g + 10 \right)^{1/4} \right] - 1.5 \right\} \le 0.6\ell_{lower,1}$
$h_{d,1}$ (m) $= \sqrt{I_s} \left\{ \left[0.42 \left(\ell_{upper,1} \right)^{1/3} \left(p_g + 0.479 \right)^{1/4} \right] - 0.457 \right\} \le 0.6\ell_{lower,1}$
$h_{d,2}$ (ft) $= \sqrt{I_s} \left\{ \left[0.43 \left(\ell_{upper,2} \right)^{1/3} \left(p_g + 10 \right)^{1/4} \right] - 1.5 \right\} \le 0.6\ell_{lower,2}$
$h_{d,2}$ (m) $= \sqrt{I_s} \left\{ \left[0.42 \left(\ell_{upper,2} \right)^{1/3} \left(p_g + 0.479 \right)^{1/4} \right] - 0.457 \right\} \le 0.6\ell_{lower,2}$
Windward drifts
$h_{d,1}$ (ft) $= 0.75\sqrt{I_s} \left\{ \left[0.43 \left(\ell_{lower,1} \right)^{1/3} \left(p_g + 10 \right)^{1/4} \right] - 1.5 \right\}$
$h_{d,1}$ (m) $= 0.75\sqrt{I_s} \left\{ \left[0.42 \left(\ell_{lower,1} \right)^{1/3} \left(p_g + 0.479 \right)^{1/4} \right] - 0.457 \right\}$
$h_{d,2}$ (ft) $= 0.75\sqrt{I_s} \left\{ \left[0.43 \left(\ell_{lower,2} \right)^{1/3} \left(p_g + 10 \right)^{1/4} \right] - 1.5 \right\}$
$h_{d,2}$ (m) $= 0.75\sqrt{I_s} \left\{ \left[0.42 \left(\ell_{lower,2} \right)^{1/3} \left(p_g + 0.479 \right)^{1/4} \right] - 0.457 \right\}$

FIGURE 3.17 Intersecting drifts at low roofs.

For parapet walls, the height of the drift is equal to three-quarters of the value of h_d determined by the equation in ASCE/SEI Figure 7.6-1 where ℓ_u is equal to the length of the roof upwind of the parapet wall.

The maximum drift load, p_d, and the total load, p_{total}, at the face of the projection or parapet wall and the width of the drift, w, depends on the clear height, h_c, and the calculated value of h_d:

- Where $h_d \leq h_c = h_{step} - h_b$:

$$p_d = \gamma h_d \text{ and } w = 4h_d$$

$$p_{total} = p_s + p_d = \gamma(h_b + h_d)$$

- Where $h_d > h_c = h_{step} - h_b$:

$$p_d = \gamma h_c \text{ and } w = 4h_d^2/h_c \leq 8h_c$$

$$p_{total} = p_s + p_d = \gamma(h_b + h_c) = \gamma h_{step}$$

In these equations, h_{step} is the height of the projection or the height of the parapet wall.

The flowchart in Fig. 3.18 can be used to determine drift loads and widths on roof projections and parapets in accordance with ASCE/SEI 7.8.

3.13   Sliding Snow

Provisions for the load caused by snow sliding off a sloped roof onto a lower roof are given in ASCE/SEI 7.9. The sliding snow load is superimposed on the balanced snow load on the lower roof and need not be used in combination with drift, unbalanced, partial, or rain-on-snow loads.

Sliding snow loads are assumed to occur on lower roofs adjacent to slippery upper roofs with slopes greater than ¼ on 12 (1.19 degrees) and not slippery roofs with slopes greater than 2 on 12 (9.46 degrees).

For adjacent structures with no horizontal separation between the two, the sliding load per length of eave is equal to $0.4p_f W$ where p_f and W are the flat roof snow load and the horizontal distance from the ridge to the eave of the upper roof, respectively. The length over which the sliding snow acts on the lower roof is 15 ft (4.6 m), which is measured horizontally from the eave of the upper roof. In cases where the length of the lower roof is less than 15 ft (4.6 m), the sliding snow load is permitted to be reduced proportionally, that is, the sliding snow can be reduced to $0.4p_f W(\ell_{lower}/15)$ [in S.I.: $0.4p_f W(\ell_{lower}/4.6)$] where ℓ_{lower} is the length of the lower roof.

The total snow load in the area subjected to sliding snow can be determined by the following for the case where $\ell_{lower} \geq 15$ ft (4.6 m):

$$p_{total} = p_s + (0.4p_f W/15) \quad (\text{lb}/\text{ft}^2) \tag{3.9}$$

$$p_{total} = p_s + (0.4p_f W/4.6) \quad (\text{kN}/\text{m}^2) \tag{3.10}$$

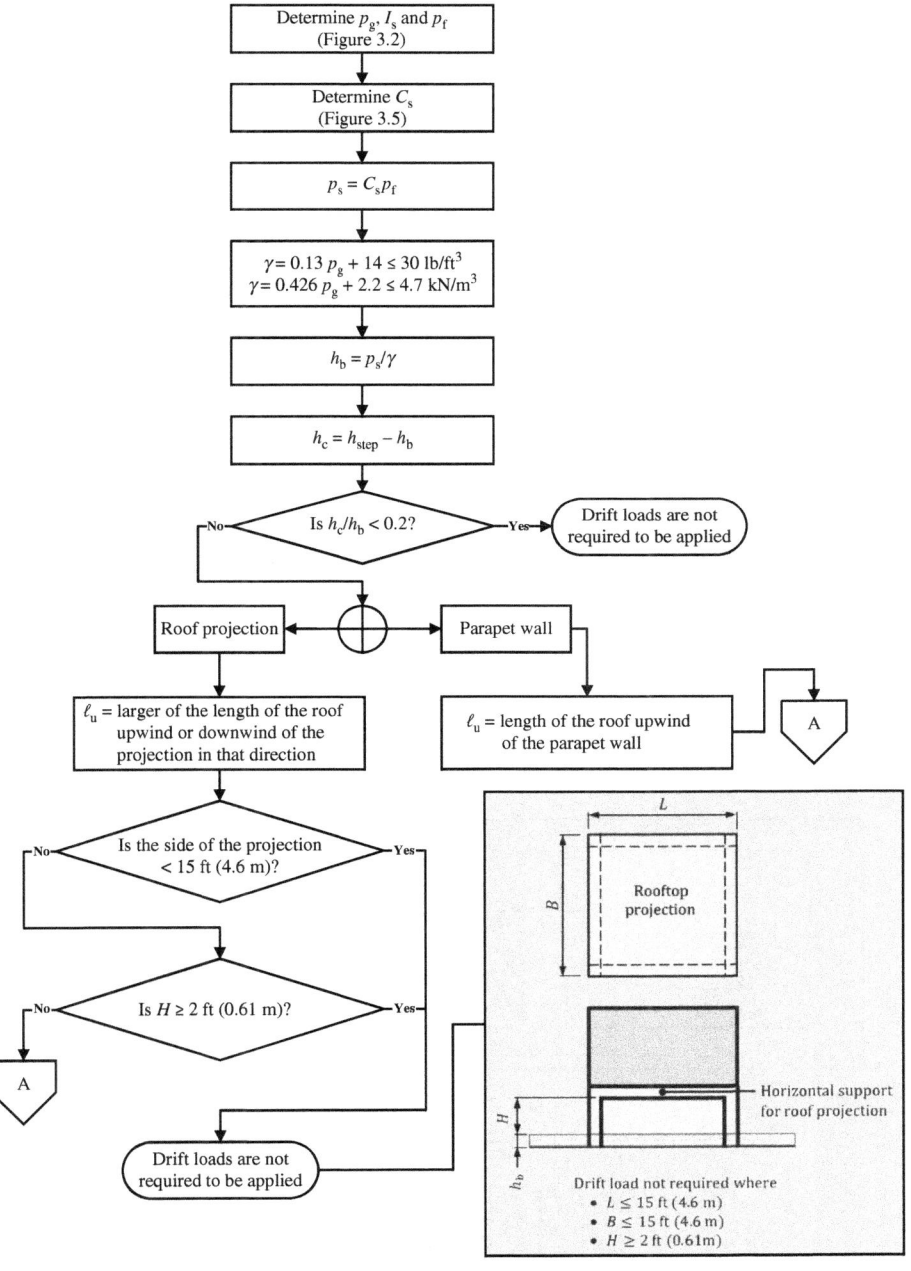

FIGURE 3.18 Flowchart to determine drift loads and widths on roof projections and parapets.

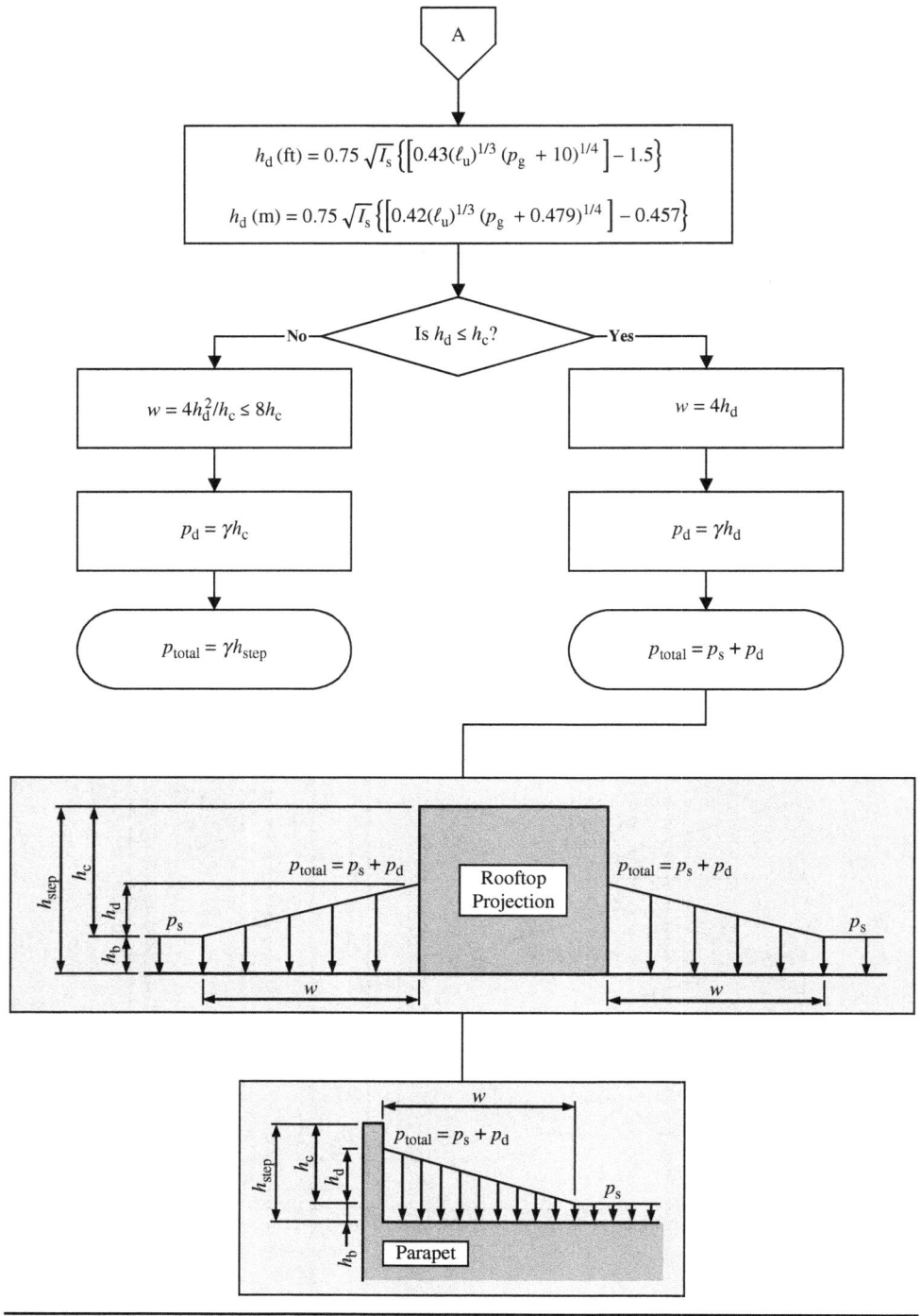

Figure 3.18   (Continued)

The corresponding total depth of snow is equal to the following:

$$h_{total} = \left[p_s + (0.4 p_f W / 15) \right] / \gamma \quad \text{(ft)} \tag{3.11}$$

$$h_{total} = \left[p_s + (0.4 p_f W / 4.6) \right] / \gamma \quad \text{(m)} \tag{3.12}$$

If h_{total} on the lower roof is greater than the vertical distance h from the top of the lower roof to the eave of the upper roof, sliding snow is blocked and a portion of the sliding snow remains on the upper roof. In such cases, the total load on the lower roof is equal to γh, which is uniformly distributed over 15 ft (4.6 m) or ℓ_{lower}, whichever is less.

Sliding snow loads on a lower roof horizontally separated by a distance s from an adjacent structure with a higher roof must be considered where $s < 15$ ft ($s < 4.6$ m) and $s < h$. The sliding snow load in this case is equal to the following:

$$p_{sliding} = 0.4 p_f W (15 - s) / 15 \quad \text{(lb/ft)} \tag{3.13}$$

$$p_{sliding} = 0.4 p_f W (4.6 - s) / 4.6 \quad \text{(kN/m)} \tag{3.14}$$

The sliding snow load is distributed over a length equal to $(15 \text{ ft} - s) [(4.6 \text{ m} - s)]$ on the lower roof, and the total snow load is determined by Eqs. (3.9) and (3.10).

The flowchart in Fig. 3.19 can be used to determine sliding snow loads in accordance with ASCE/SEI 7.9.

3.14   Rain-on-Snow Surcharge Load

A rain-on-snow surcharge load of 5 lb/ft² (0.24 kN/m²) is to be added on all roofs meeting the following two conditions (ASCE/SEI 7.10):

- The building is located where $0 < p_g \leq 20$ lb/ft² (0.96 kN/m²).
- The slope of the roof in degrees is less than $W/50$ where W is the horizontal distance in feet from the ridge to the eave [in S.I.: $W/15.2$ where W is in meters].

This surcharge load applies only to the balanced load case and need not be used in combination with drift, unbalanced minimum, or partial loads.

3.15   Ponding Instability

Where roofs do not have adequate slope or have insufficient and/or blocked drains, water due to rain or melting snow will tend to pond in low areas, which causes the roof structure to deflect. Roof drains on structures in very cold regions intermittently heated are particularly vulnerable to blockages by ice. Additional water is attracted to these low areas, leading to additional deflection. Adequate stiffness must be provided so that deflections will not continually increase until instability occurs, resulting in localized failure.

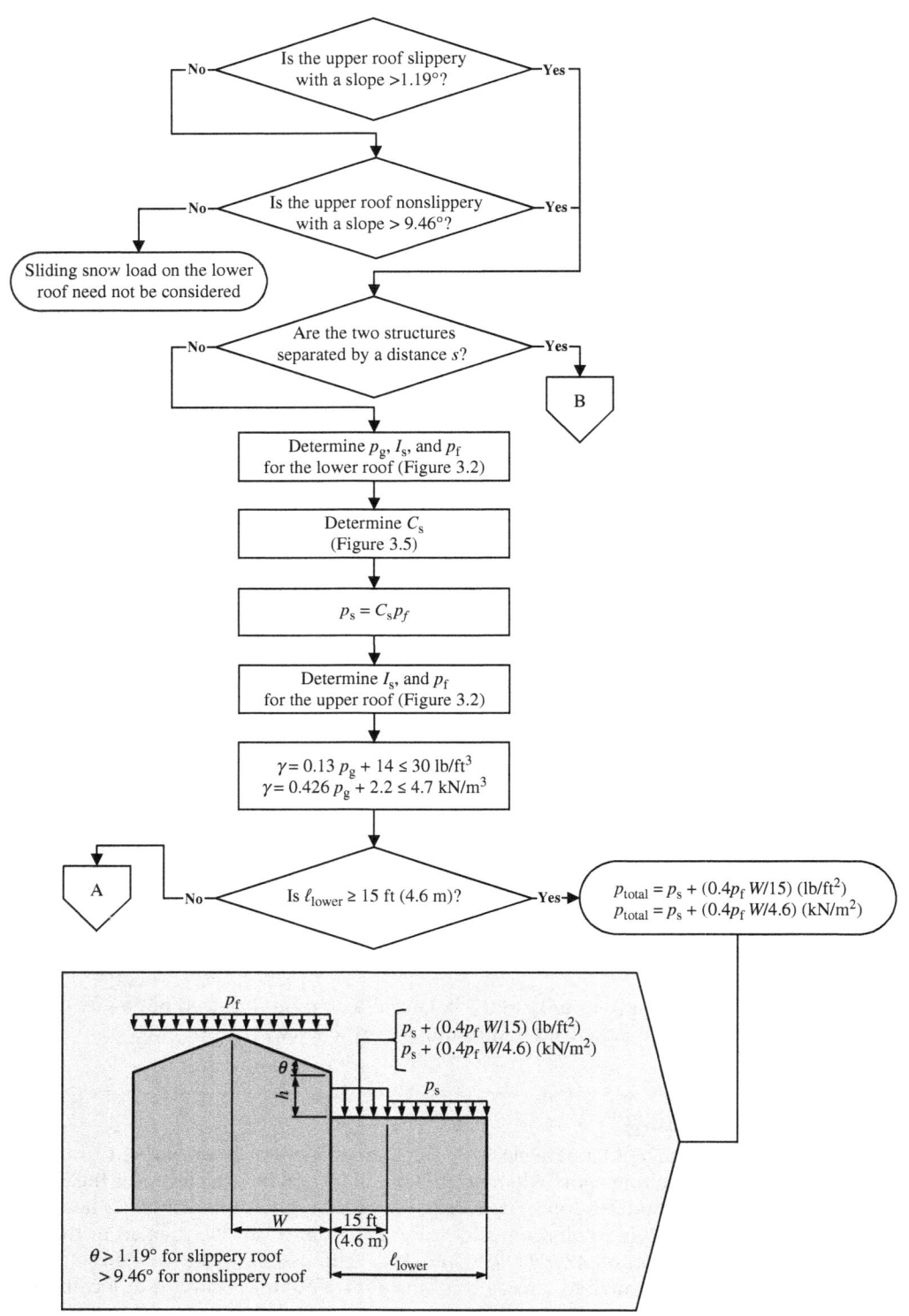

FIGURE 3.19 Flowchart to determine sliding snow loads.

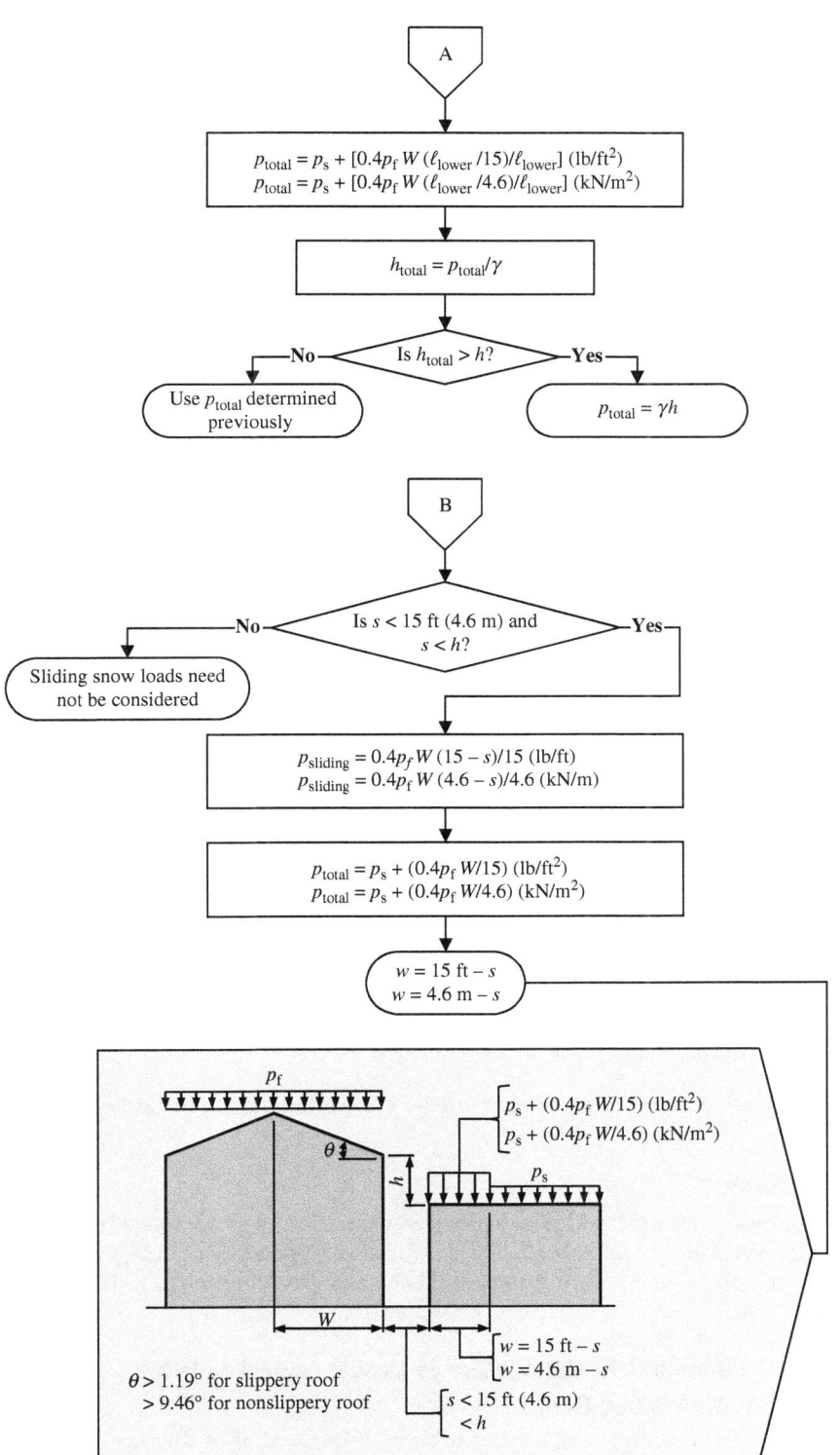

FIGURE 3.19 (Continued)

71

Provisions for ponding instability are given in ASCE/SEI 7.11. Susceptible roof bays, which are the same as those defined in ASCE/SEI 8.4 for rain loads, must be analyzed for the effects from the larger of the snow loads or the rain loads. The information in Section 2.10 of this publication related to ponding instability and loading for rain loads is also applicable to snow loads.

3.16 Existing Roofs

Requirements for increased snow loads on existing roofs due to additions and alterations are given in ASCE/SEI 7.12. Where a new structure with a higher roof is constructed within 20 ft (6.1 m) of an existing structure with a lower roof, both drift and sliding snow loads must be considered on the lower roof. The exposure of the existing roof must also be examined. For example, if the existing roof were fully exposed prior to the new building, it is likely it will be partially exposed or sheltered if the new building is taller, resulting in an increase in snow loads on the existing roof.

Another example of where snow loads may increase on an existing roof is where a new building with a gable roof is constructed alongside an existing building with a gable roof and both roofs are at the same elevation (see ASCE/SEI Figure C7.12-1). The new roof configuration is essentially a folded plate, and the valley created between the roofs will be subjected to a drift load, which did not have to be considered in the original design of the existing roof. In general, the effects of new structures or alterations on existing roofs must be carefully checked and existing roof structural members must be strengthened where required.

3.17 Snow on Open-Frame Equipment Structures

Snow loads determined in accordance with the provisions in ASCE/SEI 7.13 must be considered for all levels of an open-frame equipment structure that can retain snow. In particular, snow accumulations must be considered for the following elements:

- Flooring
- Pipes and cable trays
- Equipment and equipment platforms

The loads to be applied to each type of element are given in Table 3.6.

3.18 Examples

The following examples illustrate the determination of snow loads based on the provisions in ASCE/SEI Chapter 7. The overall design procedure in Fig. 3.1 is used as are the applicable flowcharts, figures, and tables in this chapter to determine the snow loads.

3.18.1 Example 3.1—Calculation of Design Snow Loads for a Monoslope Roof

Determine the design snow loads on the monoslope roof of the one-story building in Fig. 3.20 given the design data in Table 3.7.

	Element	Snow Loads	ASCE/SEI Figure No.
Flooring	At the top level	• Flat roof snow load, p_f, is applied over the entire level.[1],[2] • Drift loads are applied over the applicable widths.[2],[3]	7.13-1
	Below the top level[4]	Flat roof snow load, p_f, is applied over a portion of the flooring level near any open edge with a width equal to the vertical distance in elevation between the level in question and the next level above.	
Pipes and cable trays	Pipe diameter or cable tray width $\leq 0.73 p_f / \gamma$[2]	Triangular snow load of width D and height $1.37D$ with a maximum intensity equal to $1.37D\gamma$ at $(D/2)$ is applied to the pipe or cable tray where D is the pipe diameter or cable tray width.[5]	7.13-2a
	Pipe diameter or cable tray width $> 0.73 p_f / \gamma$[2]	Trapezoidal snow load of width D with a maximum intensity equal to p_f is applied to the pipe or cable tray.[5]	7.13-2b
	Clear spacing, S_p, between multiple adjacent pipes or cable trays $< (p_f / \gamma)$	Uniform cornice load of p_f is applied in the spaces between the pipes or cable trays in addition to the snow load on each individual pipe or cable tray.	7.13-3
Equipment and equipment platforms		Snow loads on the structure must include snow loads on any equipment or equipment platforms supported by the structure.[6]	—

Notes:

[1]Snow loads are applicable to flooring (grating, checkered plates, etc.) or elements that can retain snow. It is assumed open-frame members with a width greater than 8 in. (200 mm) can retain snow.

[2]Use $C_t = 1.2$ in the determination of p_f for unheated open-frame equipment structures (see Fig. 3.2 for the determination of p_f).

[3]The top level of the structure must be designed for drift loads in accordance with ASCE/SEI 7.7 and 7.9 where there are wind walls or equivalent obstructions (see Fig. 3.18 for the determination of drift loads and widths on roof projections).

[4]In cases where the top level of a structure does not have any snow-retaining surfaces, the level below the top level must be designed as the top level.

[5]The diameter or width, D, must include the thickness of insulation, where applicable.

[6]Snow accumulation on equipment and equipment platforms need not be considered where the wintertime external surface temperature is greater than 45°F (7.2°C).

TABLE 3.6 Snow Loads on Elements of Open-Frame Equipment Structures

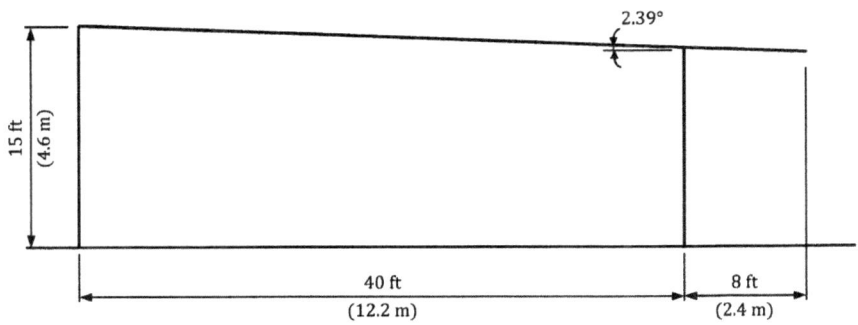

FIGURE 3.20 Elevation of the building with a monoslope roof in Example 3.1.

Location	Peoria, IL
Surface roughness	C
Occupancy	Commercial
Thermal condition	Cold, ventilated roof with $R > 25$ ft²h°F/Btu (4.4 m²K/W)
Roof exposure	Partially exposed
Roof surface	Smooth rubber membrane
Roof obstructions	None
Roof framing	Primary members spaced 20 ft (6.1 m) on center overhanging a wall and simply supported secondary members spaced 5 ft (1.5 m) on center parallel to the free draining edge of the roof

TABLE 3.7 Design Data for Example 3.1

Solution

Step 1—*Determine the ground snow load, p_g*   ASCE/SEI Figure 7.2-1

The ground snow load is equal to 20 lb/ft² (0.96 kN/m²).

Step 2—*Determine the flat roof snow load, p_f*   Fig. 3.2

- Step 2a—Determine the surface roughness category

  From the design data, the surface roughness category is given as C.

- Step 2b—Determine the exposure of the roof

  From the design data, the roof exposure is given as partially exposed.

- Step 2c—Determine the exposure factor, C_e

  Given a surface roughness category of C and a partially exposed roof exposure, $C_e = 1.0$ from ASCE/SEI Table 7.3-1.

- Step 2d—Determine the thermal factor, C_t

  From the design data, the roof is cold and ventilated with $R > 25$ ft²h°F/Btu (4.4 m²K/W). Therefore, $C_t = 1.1$ from ASCE/SEI Table 7.3-2.

- Step 2e—Determine the Risk Category of the building

  From the design data, the occupancy of the building is commercial. For a commercial occupancy, the Risk Category is II from ASCE/SEI Table 1.5-1.

- Step 2f—Determine the importance factor, I_s

  For a Risk Category II building, $I_s = 1.0$ from ASCE/SEI Table 1.5-2.

- Step 2g—Determine the flat roof snow load, p_f

$$p_f = 0.7C_eC_tI_sp_g = 0.7 \times 1.0 \times 1.1 \times 1.0 \times 20.0 = 15.4 \text{ lb/ft}^2 \tag{3.1}$$

$$p_f = 0.7C_eC_tI_sp_g = 0.7 \times 1.0 \times 1.1 \times 1.0 \times 0.96 = 0.74 \text{ kN/m}^2$$

Step 3—Determine the minimum snow load for low-slope roofs, p_m   Sec. 3.6

A minimum snow load applies to monoslope roofs with slopes less than 15 degrees (see Fig. 3.3). Because the roof slope in this example is equal to 2.39 degrees, minimum roof snow loads must be considered:

For $p_g = 20.0$ lb/ft^2, $p_m = I_s p_g = 1.0 \times 20.0 = 20.0$ lb/ft^2 Table 3.3

For $p_g = 0.96$ kN/m^2, $p_m = I_s p_g = 1.0 \times 0.96 = 0.96$ kN/m^2

Step 4—Determine the sloped roof (balanced) snow load, p_s   Sec. 3.7

- Step 4a—Determine the slope factor, C_s Fig. 3.5

It is determined in Step 2d that $C_t = 1.1$ (cold roof).

From the design data, there are no obstructions inhibiting the snow from sliding off the roof. Also, because the roof is cold, ice dams need not be considered (if an ice dam can form on a warm roof, it is considered to be an obstruction).

From the design data, the roof surface is a smooth rubber membrane. According to ASCE/SEI 7.4, smooth rubber membranes are considered to be slippery surfaces.

Because the roof is unobstructed and slippery, use the dashed line in ASCE/SEI Figure 7.4-1b to determine C_s:

For a roof slope of 2.39 degrees, which is less than 10 degrees, $C_s = 1.0$.

- Step 4b—Determine the sloped roof (balanced) snow load, p_s

$$p_s = C_s p_f = 1.0 \times 15.4 = 15.4 \text{ lb/ft}^2 \tag{3.2}$$

$$p_s = C_s p_f = 1.0 \times 0.74 = 0.74 \text{ kN/m}^2$$

Step 5—Consider loads due to ice dams   Sec. 3.8

Because the roof has been determined to be a cold roof, ice dams and the accompanying uniform load need not be considered (ASCE/SEI 7.4.5).

Step 6—Consider partial loading   Sec. 3.9

From the design data, the secondary framing members are simply supported, so partial loading need not be considered for these members (ASCE/SEI 7.5).

The primary framing members are continuous over the wall, which means partial loading must be considered for these cantilevered members. The balanced snow load determined in Step 4b is used in partial loading cases; the minimum snow load is not applicable in such cases (ASCE/SEI 7.3.4). With a center-to-center spacing of 20 ft (6.1 m), the partial loads on a typical primary member are determined as follows (see Fig. 3.8):

Case 1:

Balanced load on cantilevers $= 15.4 \times 20.0 = 308$ lb/ft

Partial load on main span = one-half of balanced load $= 0.5 \times 308 = 154$ lb/ft

Case 2:

Partial load on cantilevers = one-half of balanced load = $0.5 \times 308 = 154$ lb/ft

Balanced load on main span = 308 lb/ft

In S.I.:

Case 1:

Balanced load on cantilevers = $0.74 \times 6.1 = 4.51$ kN/m

Partial load on main span = one-half of balanced load = $0.5 \times 4.51 = 2.26$ kN/m

Case 2:

Partial load on cantilevers = one-half of balanced load = $0.5 \times 4.51 = 2.26$ kN/m

Balanced load on main span = 4.51 kN/m

Step 7—Consider unbalanced snow loads   Sec. 3.10

Because this roof is monoslope, unbalanced snow loads need not be considered (ASCE/SEI 7.6).

Step 8—Consider drifts on lower roofs   Sec. 3.11

Not applicable.

Step 9—Consider drifts on roof projections and parapets   Sec. 3.12

Not applicable.

Step 10—Consider sliding snow loads   Sec. 3.13

Not applicable.

Step 11—Consider rain-on-snow surcharge loads   Sec. 3.14

A rain-on-snow surcharge load of 5 lb/ft^2 (0.24 kN/m^2) is required for locations where (1) the ground snow load, p_g, is 20 lb/ft^2 (0.96 kN/m^2) or less and greater than zero and (2) the roof slope is less than $W/50$ (ASCE/SEI 7.10).

In this example, p_g = 20 lb/ft^2 (0.96 kN/m^2) and $W/50 = 40/50 = 0.80$ degree < roof slope = 2.39 degrees [in S.I.: $W/15.2 = 12.2/15.2 = 0.80$ degree < roof slope = 2.39 degrees]. Thus, an additional 5 lb/ft^2 (0.24 kN/m^2) need not be added to the balanced load of 15.4 lb/ft^2 (0.74 kN/m^2).

Step 12—Consider ponding instability   Sec. 3.15

Because the secondary members are parallel to the free-draining edge of the roof and the roof slope is less than 1 inch per foot (4.76 degrees), the bay is susceptible to ponding, and progressive roof deflection and ponding instability from snow meltwater must be investigated (see Fig. 2.8).

Step 13—Consider snow loads on existing roofs   Sec. 3.16

Not applicable.

Step 14—Consider snow loads on open-frame equipment structures   Sec. 3.17

Not applicable.

Step 15—Determine the snow loads on the secondary and primary roof members

It is determined in Step 6 that partial loads need not be considered for the secondary members. With a spacing of 5 ft (1.5 m), the uniform snow load on a secondary member is equal to $20.0 \times 5.0 = 100$ lb/ft [in S.I.: $0.96 \times 1.5 = 1.44$ kN/m]. The minimum snow load of 20.0 lb/ft² (0.96 kN/m²) is used to determine the uniform snow load because it is greater than the balanced snow load of 15.4 lb/ft² (0.74 kN/m²). The secondary roof members must be designed for the 100 lb/ft (1.44 kN/m) balanced snow load in combination with other applicable loads using the appropriate load combinations.

The partial loads on the primary members are determined in Step 6. With a spacing of 20 ft (6.1 m), the uniform snow load on a primary member is equal to $20.0 \times 20.0 = 400$ lb/ft [in S.I.: $0.96 \times 6.1 = 5.86$ kN/m].

The minimum and partial load cases for the primary roof members are given in Fig. 3.21. These members must be designed for these snow load cases in combination with other applicable loads using the appropriate load combinations.

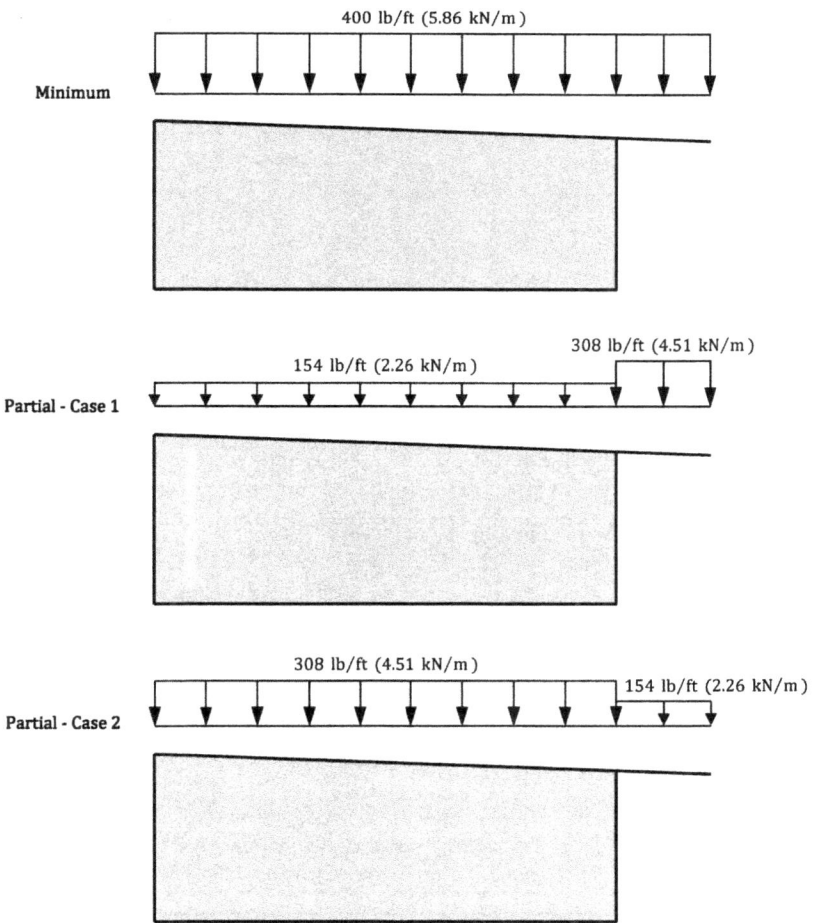

FIGURE 3.21 Minimum and partial load cases for the primary roof members in Example 3.1.

3.18.2   Example 3.2—Calculation of Design Snow Loads for a Monoslope Roof Including Ice Dams

Determine the design snow loads on the monoslope roof of the one-story building in Fig. 3.22 given the design data in Table 3.8.

Solution

Step 1—*Determine the ground snow load, p_g*   ASCE/SEI Figure 7.2-1

The ground snow load is equal to 20 lb/ft² (0.96 kN/m²).

Step 2—*Determine the flat roof snow load, p_f*   Fig. 3.2

- Step 2a—Determine the surface roughness category

  From the design data, the surface roughness category is given as C.

- Step 2b—Determine the exposure of the roof

  From the design data, the roof exposure is given as partially exposed.

- Step 2c—Determine the exposure factor, C_e

  Given a surface roughness category of C and a partially exposed roof exposure, $C_e = 1.0$ from ASCE/SEI Table 7.3-1.

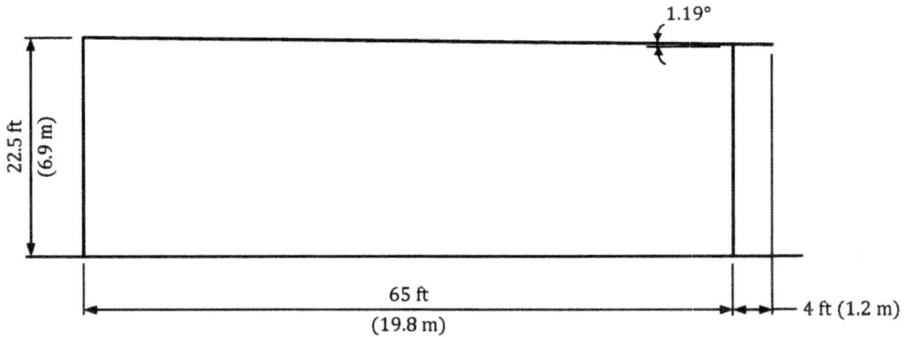

Figure 3.22   Elevation of the building with a monoslope roof in Example 3.2.

Location	Peoria, IL
Surface roughness	C
Occupancy	Essential facility (the building is required to remain operational during an emergency)
Thermal condition	Warm, ventilated roof with $R < 20$ ft²h°F/Btu (3.5 m²K/W)
Roof exposure	Partially exposed
Roof surface	Asphalt shingles
Roof obstructions	None
Roof framing	Primary members spaced 20 ft (6.1 m) on center overhanging a wall and simply supported secondary members spaced 5 ft (1.5 m) on center parallel to the free-draining edge of the roof

Table 3.8   Design Data for Example 3.2

- Step 2d—Determine the thermal factor, C_t

  From the design data, the roof is warm and ventilated roof with $R < 20$ ft²h°F/Btu (3.5 m²K/W). Because this type of thermal condition is not specifically indicated in ASCE/SEI Table 7.3-2, $C_t = 1.0$.

- Step 2e—Determine the Risk Category of the building

  From the design data, the building is an essential facility. For an essential facility, the Risk Category is IV from ASCE/SEI Table 1.5-1.

- Step 2f—Determine the importance factor, I_s

  For a Risk Category IV building, $I_s = 1.2$ from ASCE/SEI Table 1.5-2.

- Step 2g—Determine the flat roof snow load, p_f

$$p_f = 0.7 C_e C_t I_s p_g = 0.7 \times 1.0 \times 1.0 \times 1.2 \times 20.0 = 16.8 \text{ lb/ft}^2 \qquad (3.1)$$

$$p_f = 0.7 C_e C_t I_s p_g = 0.7 \times 1.0 \times 1.0 \times 1.2 \times 0.96 = 0.81 \text{ kN/m}^2$$

Step 3—Determine the minimum snow load for low-slope roofs, p_m   Sec. 3.6

A minimum snow load applies to monoslope roofs with slopes less than 15 degrees (see Fig. 3.3). Because the roof slope in this example is equal to 1.19 degrees, minimum roof snow loads must be considered:

$$\text{For } p_g = 20.0 \text{ lb/ft}^2, \ p_m = I_s p_g = 1.2 \times 20.0 = 24.0 \text{ lb/ft}^2 \qquad \text{Table 3.3}$$

$$\text{For } p_g = 0.96 \text{ kN/m}^2, \ p_m = I_s p_g = 1.2 \times 0.96 = 1.15 \text{ kN/m}^2$$

Step 4—Determine the sloped roof (balanced) snow load, p_s   Sec. 3.7

- Step 4a—Determine the slope factor, C_s Fig. 3.5

  It is determined in Step 2d that $C_t = 1.0$ (warm roof).

  The roof is warm and ventilated with $R < 20$ ft²h°F/Btu (3.5 m²K/W), which means it is possible for an ice dam to form at the eave (ASCE/SEI 7.4.5). An ice dam is considered to be an obstruction because it can prevent snow from sliding off the roof.

  From the design data, the roof surface consists of asphalt shingles. According to ASCE/SEI 7.4, asphalt shingles are not considered to be slippery surfaces.

  Because the roof is obstructed and not slippery, use the solid line in ASCE/SEI Figure 7.4-1a to determine C_s:

  For a roof slope of 1.19 degrees, which is less than 30 degrees, $C_s = 1.0$.

- Step 4b—Determine the sloped roof (balanced) snow load, p_s

$$p_s = C_s p_f = 1.0 \times 16.8 = 16.8 \text{ lb/ft}^2 \qquad (3.2)$$

$$p_s = C_s p_f = 1.0 \times 0.81 = 0.81 \text{ kN/m}^2$$

Step 5—Consider loads due to ice dams   Sec. 3.8

Because the roof is warm and ventilated with $R < 20$ ft²h°F/Btu (3.5 m²K/W), it is possible for an ice dam to form at the eave (ASCE/SEI 7.4.5).

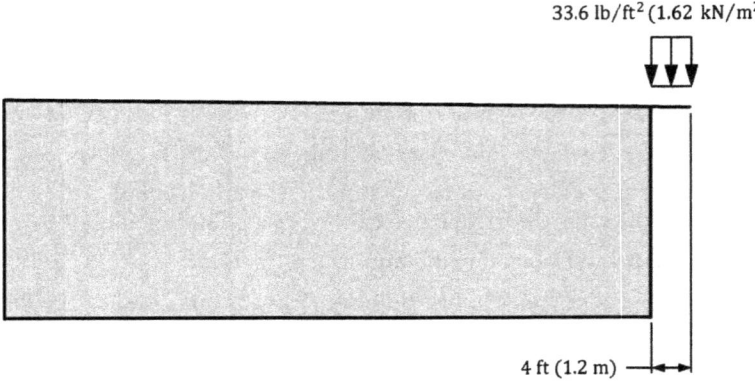

33.6 lb/ft² (1.62 kN/m²)

4 ft (1.2 m)

FIGURE 3.23 Uniform ice dam load on the roof overhang of the building in Example 3.2.

A uniformly distributed load of $2p_f = 2 \times 16.8 = 33.6$ lb/ft² [in S.I.: $2p_f = 2 \times 0.81 = 1.62$ kN/m²] must be applied on the 4-ft (1.22-m) overhang (see Fig. 3.23). Only the dead load must be present on the roof when this load is applied.

Step 6—Consider partial loading   Sec. 3.9

From the design data, the secondary framing members are simply supported, so partial loading need not be considered for these members (ASCE/SEI 7.5).

The primary framing members are continuous over the wall, which means partial loading must be considered for these cantilevered members. The balanced snow load determined in Step 4b is used in partial loading cases; the minimum snow load is not applicable in such cases (ASCE/SEI 7.3.4). With a center-to-center spacing of 20 ft (6.1 m), the partial loads on a typical primary member are determined as follows (see Fig. 3.8):

Case 1:

Balanced load on cantilevers = $16.8 \times 20.0 = 336$ lb/ft

Partial load on main span = one-half of balanced load = $0.5 \times 336 = 168$ lb/ft

Case 2:

Partial load on cantilevers = one-half of balanced load = $0.5 \times 336 = 168$ lb/ft

Balanced load on main span = 336 lb/ft

In S.I.:

Case 1:

Balanced load on cantilevers = $0.81 \times 6.1 = 4.94$ kN/m

Partial load on main span = one-half of balanced load = $0.5 \times 4.94 = 2.47$ kN/m

Case 2:

Partial load on cantilevers = one-half of balanced load = $0.5 \times 4.94 = 2.47$ kN/m

Balanced load on main span = 4.94 kN/m

Step 7—Consider unbalanced snow loads   Sec. 3.10

Because this roof is monoslope, unbalanced snow loads need not be considered (ASCE/SEI 7.6).

Step 8—Consider drifts on lower roofs   Sec. 3.11

Not applicable.

Step 9—Consider drifts on roof projections and parapets   Sec. 3.12

Not applicable.

Step 10—Consider sliding snow loads   Sec. 3.13

Not applicable.

Step 11—Consider rain-on-snow surcharge loads   Sec. 3.14

A rain-on-snow surcharge load of 5 lb/ft² (0.24 kN/m²) is required for locations where (1) the ground snow load, p_g, is 20 lb/ft² (0.96 kN/m²) or less and greater than zero and (2) the roof slope is less than $W/50$ (ASCE/SEI 7.10).

In this example, $p_g = 20$ lb/ft² (0.96 kN/m²) and $W/50 = 65/50 = 1.30$ degrees > roof slope = 1.19 degrees [in S.I.: $W/15.2 = 19.8/15.2 = 1.30$ degrees > roof slope = 1.19 degrees]. Thus, an additional 5 lb/ft² (0.24 kN/m²) must be added to the balanced load of 16.8 lb/ft² (0.81 kN/m²).

Step 12—Consider ponding instability   Sec. 3.15

Because the secondary members are parallel to the free-draining edge of the roof and the roof slope is less than 1 inch per foot (4.76 degrees), the bay is susceptible to ponding, and progressive roof deflection and ponding instability from snow meltwater must be investigated (see Fig. 2.8).

Step 13—Consider snow loads on existing roofs   Sec. 3.16

Not applicable.

Step 14—Consider snow loads on open-frame equipment structures   Sec. 3.17

Not applicable.

Step 15—Determine the snow loads on the secondary and primary roof members

It is determined in Step 6 that partial loads need not be considered for the secondary members. With a spacing of 5 ft (1.5 m), the uniform snow load on a secondary member is equal to $24.0 \times 5.0 = 120$ lb/ft [in S.I.: $1.15 \times 1.5 = 1.73$ kN/m]. The minimum snow load of 24.0 lb/ft² (1.15 kN/m²) is used to determine the total uniform snow load because it is greater than the balanced snow load of $16.8 + 5.0 = 21.8$ lb/ft² [in S.I.: $0.81 + 0.24 = 1.05$ kN/m²]. The secondary roof members must be designed for the 120 lb/ft (1.73 kN/m) balanced snow load in combination with other applicable loads using the appropriate load combinations.

The partial loads on the primary members are determined in Step 6. With a spacing of 20 ft (6.1 m), the uniform snow load on a primary member is equal to $24.0 \times 20.0 = 480$ lb/ft [in S.I.: $1.15 \times 6.1 = 7.02$ kN/m].

The minimum and partial load cases for the primary roof members are given in Fig. 3.24. These members must be designed for these snow load cases in combination with other applicable loads using the appropriate load combinations.

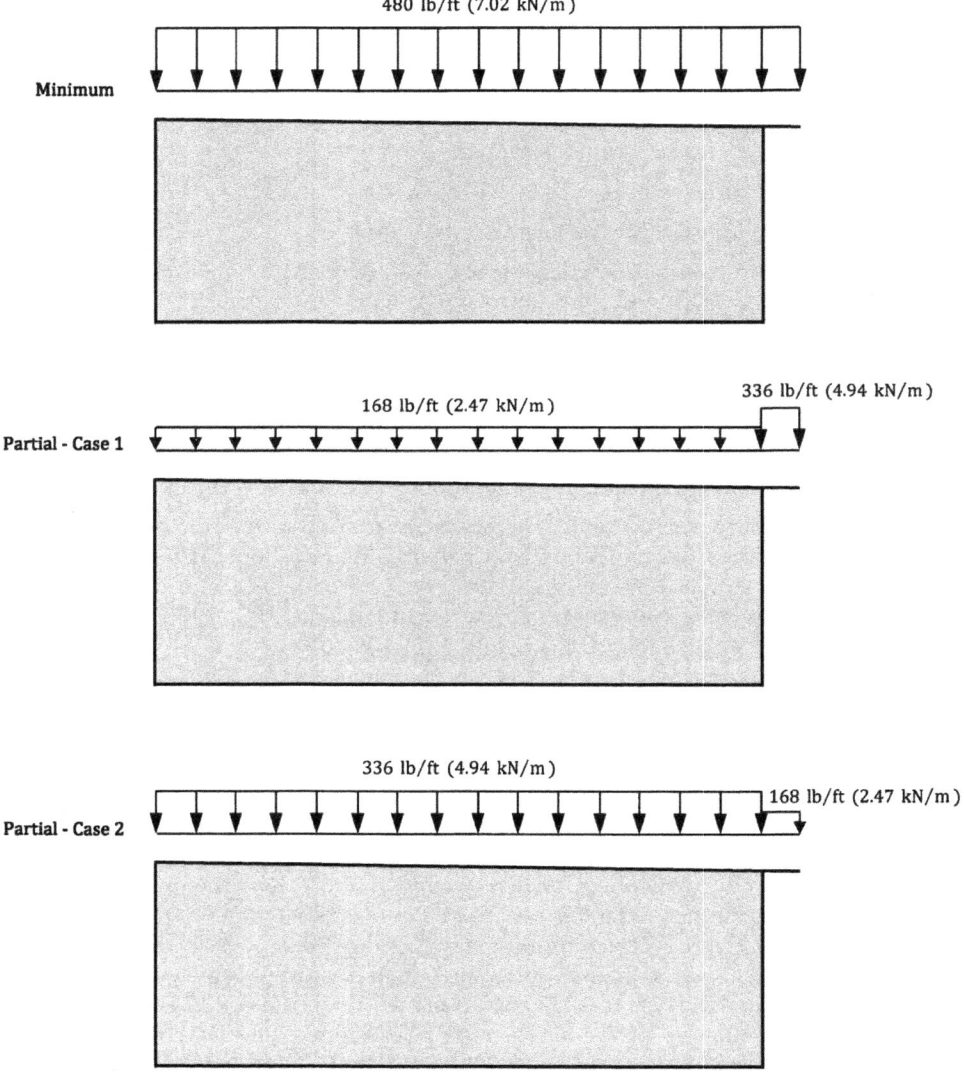

FIGURE 3.24   Minimum and partial load cases for the primary roof members in Example 3.2.

3.18.3 Example 3.3—Calculation of Design Snow Loads for a Gable Roof (Roof Slope = 1.19 degrees)

Determine the design snow loads on the gable roof of the one-story building in Fig. 3.25 given the design data in Table 3.9.

Solution

Step 1—Determine the ground snow load, p_g ASCE/SEI Table 7.2-2

The ground snow load is equal to 40 lb/ft² (1.92 kN/m²).

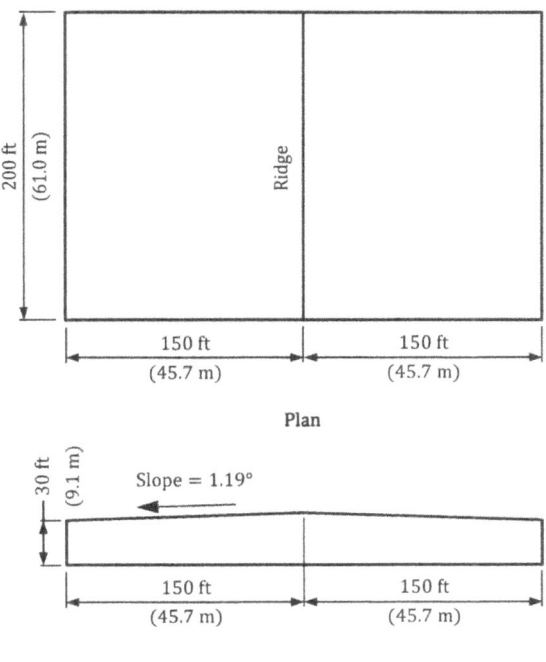

Plan

Elevation

Figure 3.25 Plan and elevation of the building with a gable roof in Example 3.3.

Location	Boulder, CO
Surface roughness	C
Occupancy	Storage (nonhazardous and nontoxic materials)
Thermal condition	Unheated
Roof exposure	Fully exposed
Roof surface	Smooth bituminous membrane
Roof obstructions	None
Roof framing	Simply supported primary and secondary members with the secondary members spanning parallel to the free-draining edges of the roof

Table 3.9 Design Data for Example 3.3

Step 2—Determine the flat roof snow load, p_f Fig. 3.2

- Step 2a—Determine the surface roughness category

  From the design data, the surface roughness category is given as C.

- Step 2b—Determine the exposure of the roof

  From the design data, the roof exposure is given as fully exposed.

- Step 2c—Determine the exposure factor, C_e

  Given a surface roughness category of C and a fully exposed roof exposure, $C_e = 0.9$ from ASCE/SEI Table 7.3-1.

- Step 2d—Determine the thermal factor, C_t

  From the design data, the building is unheated. Therefore, $C_t = 1.2$ from ASCE/SEI Table 7.3-2.

- Step 2e—Determine the Risk Category of the building

  From the design data, the building is used to store nonhazardous and nontoxic materials. Therefore, the Risk Category is II from ASCE/SEI Table 1.5-1.

- Step 2f—Determine the importance factor, I_s

  For a Risk Category II building, $I_s = 1.0$ from ASCE/SEI Table 1.5-2.

- Step 2g—Determine the flat roof snow load, p_f

$$p_f = 0.7C_eC_tI_sp_g = 0.7 \times 0.9 \times 1.2 \times 1.0 \times 40.0 = 30.2 \text{ lb/ft}^2 \qquad (3.1)$$

$$p_f = 0.7C_eC_tI_sp_g = 0.7 \times 0.9 \times 1.2 \times 1.0 \times 1.92 = 1.45 \text{ kN/m}^2$$

Step 3—Determine the minimum snow load for low-slope roofs, p_m Sec. 3.6

A minimum snow load applies to gable roofs with slopes less than 15 degrees (see Fig. 3.3). Because the roof slope in this example is equal to 1.19 degrees, minimum roof snow loads must be considered:

For $p_g = 40.0 \text{ lb/ft}^2 > 20.0 \text{ lb/ft}^2$, $p_m = 20I_s = 20.0 \times 1.0 = 20.0 \text{ lb/ft}^2$ Table 3.3

For $p_g = 1.92 \text{ kN/m}^2 > 0.96 \text{ kN/m}^2$, $p_m = 0.96I_s = 0.96 \times 1.0 = 0.96 \text{ kN/m}^2$

Step 4—Determine the sloped roof (balanced) snow load, p_s Sec. 3.7

- Step 4a—Determine the slope factor, C_s Fig. 3.5

  It is determined in Step 2d that $C_t = 1.2$ (cold roof).

  From the design data, there are no obstructions inhibiting the snow from sliding off the roof. Also, because the roof is cold, ice dams need not be considered (if an ice dam can form on a warm roof, it is considered to be an obstruction).

  From the design data, the roof surface is a smooth bituminous membrane. According to ASCE/SEI 7.4, smooth bituminous membranes are considered to be slippery surfaces.

  Because the roof is unobstructed and slippery, use the dashed line in ASCE/SEI Figure 7.4-1c to determine C_s:

  For a roof slope of 1.19 degrees, which is less than 15 degrees, $C_s = 1.0$.

• Step 4b—Determine the sloped roof (balanced) snow load, p_s Fig. 3.9

$$p_s = C_s p_f = 1.0 \times 30.2 = 30.2 \text{ lb/ft}^2 \tag{3.2}$$

$$p_s = C_s p_f = 1.0 \times 1.45 = 1.45 \text{ kN/m}^2$$

Step 5—Consider loads due to ice dams   Sec. 3.8

Because the roof has been determined to be a cold roof, ice dams and the accompanying uniform load need not be considered (ASCE/SEI 7.4.5).

Step 6—Consider partial loading   Sec. 3.9

From the design data, the primary and secondary framing members are simply supported, so partial loading need not be considered for these members (ASCE/SEI 7.5).

Step 7—Consider unbalanced snow loads   Sec. 3.10

Unbalanced snow loads need not be considered for gable roofs with a slope exceeding 30.3 degrees or with a slope less than 2.39 degrees (ASCE/SEI 7.6.1). The roof slope in this example is 1.19 degrees, which is less than 2.39 degrees, so unbalanced loads are not required to be applied.

Step 8—Consider drifts on lower roofs   Sec. 3.11

Not applicable.

Step 9—Consider drifts on roof projections and parapets   Sec. 3.12

Not applicable.

Step 10—Consider sliding snow loads   Sec. 3.13

Not applicable.

Step 11—Consider rain-on-snow surcharge loads   Sec. 3.14

A rain-on-snow surcharge load of 5 lb/ft² (0.24 kN/m²) is required for locations where (1) the ground snow load, p_g, is 20 lb/ft² (0.96 kN/m²) or less and greater than zero and (2) the roof slope is less than $W/50$ (ASCE/SEI 7.10).

In this example, p_g = 40 lb/ft² (1.92 kN/m²), which is greater than 20 lb/ft² (0.96 kN/m²), so a rain-on-snow load is not required.

Step 12—Consider ponding instability   Sec. 3.15

Because the secondary members are parallel to the free-draining edges of the roof and the roof slope is less than 1 inch per foot (4.76 degrees), the bay is susceptible to ponding, and progressive roof deflection and ponding instability from snow meltwater must be investigated (see Fig. 2.8).

Step 13—Consider snow loads on existing roofs   Sec. 3.16

Not applicable.

Step 14—Consider snow loads on open-frame equipment structures   Sec. 3.17

Not applicable.

The balanced snow load on the roof of the building in this example is depicted in Fig. 3.26.

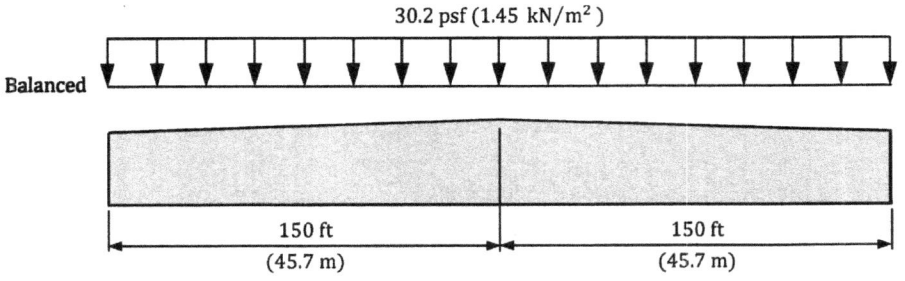

Figure 3.26   Balanced snow load for the building in Example 3.3.

3.18.4   Example 3.4–Calculation of Design Snow Loads for a Gable Roof (Roof Slope = 2.39 degrees)

Determine the design snow loads for the one-story building in Fig. 3.25 with a roof slope of 2.39 degrees. Use the design data in Table 3.9.

Solution

Steps 1 through 6—Results are the same as in Example 3.3

$$p_g = 40 \text{ lb/ft}^2 \text{ (1.92 kN/m}^2)$$
$$C_e = 0.9$$
$$C_t = 1.2$$
$$I_s = 1.0$$
$$p_f = p_s = 30.2 \text{ lb/ft}^2 \text{ (1.45 kN/m}^2)$$
$$p_m = 20.0 \text{ lb/ft}^2 \text{ (0.96 kN/m}^2)$$

Ice dam loads and partial loading need not be considered.

Step 7—Consider unbalanced snow loads   Sec. 3.10

Unbalanced snow loads need not be considered for gable roofs with a slope exceeding 30.3 degrees or with a slope less than 2.39 degrees (ASCE/SEI 7.6.1). The roof slope in this example is 2.39 degrees, which means unbalanced loads must be considered.

Because $W = 150$ ft (45.7 m) > 20 ft (6.1 m), the unbalanced load consists of the following (see Fig. 3.9):

- Windward side

  Unbalanced load = $0.3p_s = 0.3 \times 30.2 = 9.1$ lb/ft² is applied over the entire length of the windward side.

- Leeward side

$$h_d = \sqrt{I_s}\left\{[0.43(W)^{1/3}(p_g + 10)^{1/4}] - 1.5\right\} \qquad (3.3)$$
$$= \sqrt{1.0} \times \left\{[0.43 \times (150.0)^{1/3} \times (40.0 + 10)^{1/4}] - 1.5\right\} = 4.6 \text{ ft}$$

$$\gamma = 0.13p_g + 14 = (0.13 \times 40.0) + 14 = 19.2 \text{ lb/ft}^3 < 30.0 \text{ lb/ft}^3 \qquad (3.5)$$

$$S = \text{roof slope run for a rise of one} = 1/\tan 2.39° = 24$$

Balanced load $= p_s = 30.2$ lb/ft^2 is applied over the entire length of the leeward side.

Uniform pressure of $h_d\gamma/\sqrt{S} = (4.6 \times 19.2)/\sqrt{24} = 18.0$ lb/ft^2 is applied from the ridge a distance of $8h_d\sqrt{S}/3 = (8 \times 4.6 \times \sqrt{24})/3 = 60.1$ ft.

In S.I.:

- Windward side

Unbalanced load $= 0.3p_s = 0.3 \times 1.45 = 0.44$ kN/m^2 is applied over the entire length of the windward side.

- Leeward side

$$h_d = \sqrt{I_s}\left\{[0.42(W)^{1/3}(p_g + 0.479)^{1/4}] - 0.457\right\} \tag{3.4}$$

$$= \sqrt{1.0} \times \left\{[0.42 \times (45.7)^{1/3} \times (1.92 + 0.479)^{1/4}] - 0.457\right\} = 1.4 \text{ m}$$

$$\gamma = 0.426p_g + 2.2 = (0.426 \times 1.92) + 2.2 = 3.0 \text{ kN/m}^3 < 4.7 \text{ kN/m}^3 \tag{3.6}$$

$$S = \text{roof slope run for a rise of one} = 1/\tan 2.39° = 24$$

Balanced load $= p_s = 1.45$ kN/m^2 is applied over the entire length of the leeward side.

Uniform pressure of $h_d\gamma/\sqrt{S} = (1.4 \times 3.0)/\sqrt{24} = 0.86$ kN/m^2 is applied from the ridge a distance of $8h_d\sqrt{S}/3 = (8 \times 1.4 \times \sqrt{24})/3 = 18.3$ m.

Step 8—Consider drifts on lower roofs   Sec. 3.11

Not applicable.

Step 9—Consider drifts on roof projections and parapets   Sec. 3.12

Not applicable.

Step 10—Consider sliding snow loads   Sec. 3.13

Not applicable.

Step 11—Consider rain-on-snow surcharge loads   Sec. 3.14

A rain-on-snow surcharge load of 5 lb/ft^2 (0.24 kN/m^2) is required for locations where (1) the ground snow load, p_g, is 20 lb/ft^2 (0.96 kN/m^2) or less and greater than zero and (2) the roof slope is less than $W/50$ (ASCE/SEI 7.10).

In this example, $p_g = 40$ lb/ft^2 (1.92 kN/m^2), which is greater than 20 lb/ft^2 (0.96 kN/m^2), so a rain-on-snow load is not required.

Step 12—Consider ponding instability   Sec. 3.15

Because the secondary members are parallel to the free-draining edges of the roof and the roof slope is less than 1 inch per foot (4.76 degrees), the bay is susceptible to ponding, and progressive roof deflection and ponding instability from snow meltwater must be investigated (see Fig. 2.8).

Step 13—Consider snow loads on existing roofs   Sec. 3.16

Not applicable.

Step 14—Consider snow loads on open-frame equipment structures   Sec. 3.17

Not applicable.

The balanced and unbalanced snow loads on the roof of the building in this example are depicted in Fig. 3.27.

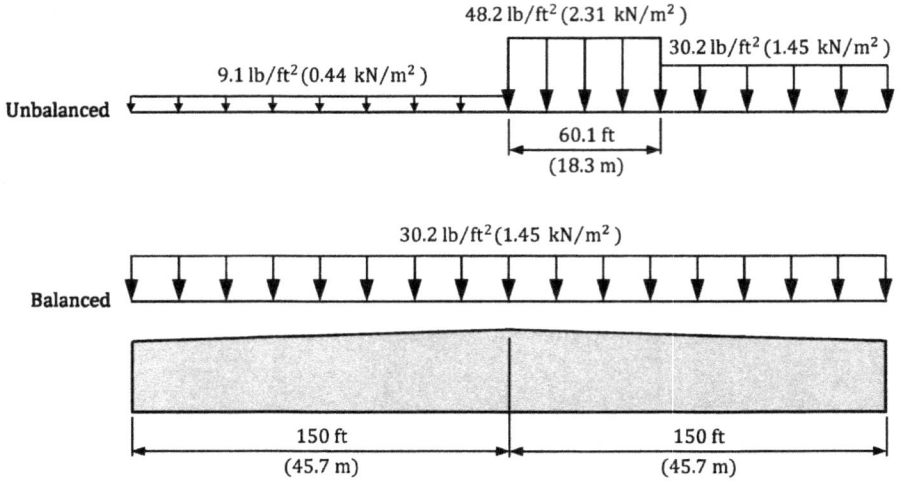

Figure 3.27 Balanced and unbalanced snow loads for the building in Example 3.4.

3.18.5 Example 3.5—Calculation of Design Snow Loads for a Gable Roof (Roof Slope = 16.7 degrees)

Determine the design snow loads for the one-story utility building in Fig. 3.28 given the design data in Table 3.10.

Solution

Step 1—Determine the ground snow load, p_g   ASCE/SEI Table 7.2-3

The ground snow load is equal to 31 lb/ft² (1.48 kN/m²).

Step 2—Determine the flat roof snow load, p_f   Fig. 3.2

- Step 2a—Determine the surface roughness category

  From the design data, the surface roughness category is given as C.

- Step 2b—Determine the exposure of the roof

  From the design data, the roof exposure is given as sheltered.

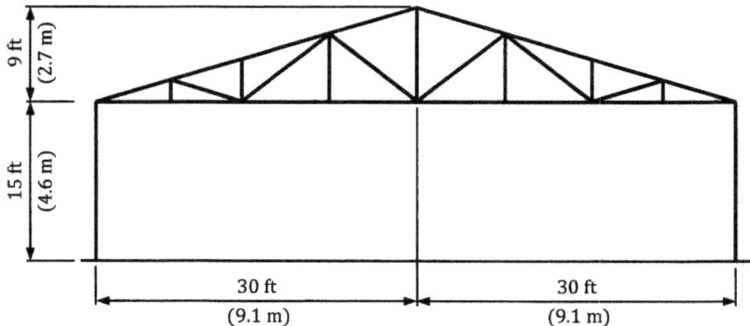

Figure 3.28 Elevation of the building with a gable roof in Example 3.5.

Location	Pocatello, ID
Surface roughness	C
Occupancy	Utility
Thermal condition	Open-air structure (no walls)
Roof exposure	Sheltered
Roof surface	Wood shingles
Roof obstructions	None
Roof framing	Trusses spaced 5 ft (1.5 m) on center

TABLE 3.10 Design Data for Example 3.5

- Step 2c—Determine the exposure factor, C_e

  Given a surface roughness category of C and a sheltered roof exposure, $C_e = 1.1$ from ASCE/SEI Table 7.3-1.

- Step 2d—Determine the thermal factor, C_t

  From the design data, the building is an open-air structure. Therefore, $C_t = 1.2$ from ASCE/SEI Table 7.3-2.

- Step 2e—Determine the Risk Category of the building

  From the design data, the building is a utility building, which represents a low risk to human life in the event of failure. Therefore, the Risk Category is I from ASCE/SEI Table 1.5-1.

- Step 2f—Determine the importance factor, I_s

  For a Risk Category I building, $I_s = 0.8$ from ASCE/SEI Table 1.5-2.

- Step 2g—Determine the flat roof snow load, p_f

$$p_f = 0.7C_eC_tI_sp_g = 0.7 \times 1.1 \times 1.2 \times 0.8 \times 31.0 = 22.9 \text{ lb/ft}^2 \qquad (3.1)$$

$$p_f = 0.7C_eC_tI_sp_g = 0.7 \times 1.1 \times 1.2 \times 0.8 \times 1.48 = 1.09 \text{ kN/m}^2$$

Step 3—Determine the minimum snow load for low-slope roofs, p_m   Sec. 3.6

A minimum snow load applies to gable roofs with slopes less than 15 degrees (see Fig. 3.3). Because the roof slope in this example is equal to $\tan^{-1}(9/30) = 16.7$ degrees, minimum roof snow loads do not apply.

Step 4—Determine the sloped roof (balanced) snow load, p_s   Sec. 3.7

- Step 4a—Determine the slope factor, C_s Fig. 3.5

  It is determined in Step 2d that $C_t = 1.2$ (cold roof).

    From the design data, there are no obstructions inhibiting the snow from sliding off the roof. Also, because the roof is cold, ice dams need not be considered (if an ice dam can form on a warm roof, it is considered to be an obstruction).

    From the design data, the roof surface has wood shingles. According to ASCE/SEI 7.4, wood shingles are not considered to be slippery.

Because the roof is unobstructed and not slippery, use the solid line in ASCE/SEI Figure 7.4-1c to determine C_s:

For a roof slope of 16.7 degrees, which is less than 45 degrees, $C_s = 1.0$.

- Step 4b—Determine the sloped roof (balanced) snow load, p_s Fig. 3.9

$$p_s = C_s p_f = 1.0 \times 22.9 = 22.9 \text{ lb/ft}^2 \tag{3.2}$$
$$p_s = C_s p_f = 1.0 \times 1.09 = 1.09 \text{ kN/m}^2$$

Step 5—Consider loads due to ice dams   Sec. 3.8

Because the roof has been determined to be a cold roof, ice dams and the accompanying uniform load need not be considered (ASCE/SEI 7.4.5).

Step 6—Consider partial loading   Sec. 3.9

Partial loads need not be applied to structural members spanning perpendicular to the ridgeline in gable roofs with slopes between 2.39 degrees and 30.3 degrees (ASCE/SEI 7.5).

Because the roof slope of 16.7 degrees is between these limits, partial loading need not be considered (note: partial loads on individual members of roof trusses are generally not considered).

Step 7—Consider unbalanced snow loads   Sec. 3.10

Unbalanced snow loads need not be considered for gable roofs with a slope exceeding 30.3 degrees or with a slope less than 2.39 degrees (ASCE/SEI 7.6.1). The roof slope in this example is 16.7 degrees, which means unbalanced loads must be considered.

Because $W = 30$ ft (9.1 m) > 20 ft (6.1 m), the unbalanced load consists of the following (see Fig. 3.9):

- Windward side

Unbalanced load $= 0.3 p_s = 0.3 \times 22.9 = 6.9 \text{ lb/ft}^2$ is applied over the entire length of the windward side.

Unbalanced load on truss $= 6.9 \times 5.0 = 34.5 \text{ lb/ft}$

- Leeward side

$$h_d = \sqrt{I_s} \left\{ [0.43(W)^{1/3}(p_g + 10)^{1/4}] - 1.5 \right\} \tag{3.3}$$
$$= \sqrt{0.8} \times \left\{ [0.43 \times (30.0)^{1/3} \times (31.0 + 10)^{1/4}] - 1.5 \right\} = 1.7 \text{ ft}$$

$$\gamma = 0.13 p_g + 14 = (0.13 \times 31.0) + 14 = 18.0 \text{ lb/ft}^3 < 30.0 \text{ lb/ft}^3 \tag{3.5}$$

$$S = \text{roof slope run for a rise of one} = 1/\tan 16.7° = 3.3$$

Balanced load $= p_s = 22.9 \text{ lb/ft}^2$ is applied over the entire length of the leeward side.

Uniform pressure of $h_d \gamma / \sqrt{S} = (1.7 \times 18.0)/\sqrt{3.3} = 16.8 \text{ lb/ft}^2$ is applied from the ridge a distance of $8 h_d \sqrt{S}/3 = (8 \times 1.7 \times \sqrt{3.3})/3 = 8.2$ ft.

Unbalanced load on truss over the 8.2-ft length $= (22.9 + 16.8) \times 5.0 = 198.5 \text{ lb/ft}$

Unbalanced load on truss over the remaining length $= 22.9 \times 5.0 = 114.5 \text{ lb/ft}$

In S.I.:

- Windward side

  Unbalanced load $= 0.3 p_s = 0.3 \times 1.09 = 0.33$ kN/m^2 is applied over the entire length of the windward side.
  Unbalanced load on truss $= 0.33 \times 1.5 = 0.50$ kN/m

- Leeward side

$$h_d = \sqrt{I_s} \left\{ [0.42(W)^{1/3}(p_g + 0.479)^{1/4}] - 0.457 \right\} \tag{3.4}$$

$$= \sqrt{0.8} \times \left\{ [0.42 \times (9.1)^{1/3} \times (1.48 + 0.479)^{1/4}] - 0.457 \right\} = 0.52 \text{ m}$$

$$\gamma = 0.426 p_g + 2.2 = (0.426 \times 1.48) + 2.2 = 2.8 \text{ kN/m}^3 < 4.7 \text{ kN/m}^3 \tag{3.6}$$

$$S = \text{roof slope run for a rise of one} = 1/\tan 16.7° = 3.3$$

Balanced load $= p_s = 1.09$ kN/m^2 is applied over the entire length of the leeward side.
Uniform pressure of $h_d \gamma / \sqrt{S} = (0.52 \times 2.8)/\sqrt{3.3} = 0.80$ kN/m^2 is applied from the ridge a distance of $8 h_d \sqrt{S}/3 = (8 \times 0.52 \times \sqrt{3.3})/3 = 2.5$ m.

Unbalanced load on truss over the 2.5-m length $= (1.09 + 0.80) \times 1.5 = 2.84$ kN/m

Unbalanced load on truss over the remaining length $= 1.09 \times 1.5 = 1.64$ kN/m

Step 8—Consider drifts on lower roofs   Sec. 3.11

Not applicable.

Step 9—Consider drifts on roof projections and parapets   Sec. 3.12

Not applicable.

Step 10—Consider sliding snow loads   Sec. 3.13

Not applicable.

Step 11—Consider rain-on-snow surcharge loads   Sec. 3.14

A rain-on-snow surcharge load of 5 lb/ft^2 (0.24 kN/m^2) is required for locations where (1) the ground snow load, p_g, is 20 lb/ft^2 (0.96 kN/m^2) or less and greater than zero and (2) the roof slope is less than $W/50$ (ASCE/SEI 7.10).
In this example, $p_g = 31$ lb/ft^2 (1.48 kN/m^2), which is greater than 20 lb/ft^2 (0.96 kN/m^2), so a rain-on-snow load is not required.

Step 12—Consider ponding instability   Sec. 3.15

Because the roof slope is greater than 1 inch per foot (4.76 degrees), the bay is not susceptible to ponding, and progressive roof deflection and ponding instability from snow meltwater need not be investigated (see ASCE/SEI 7.11 and 8.4).

Step 13—Consider snow loads on existing roofs   Sec. 3.16

Not applicable.

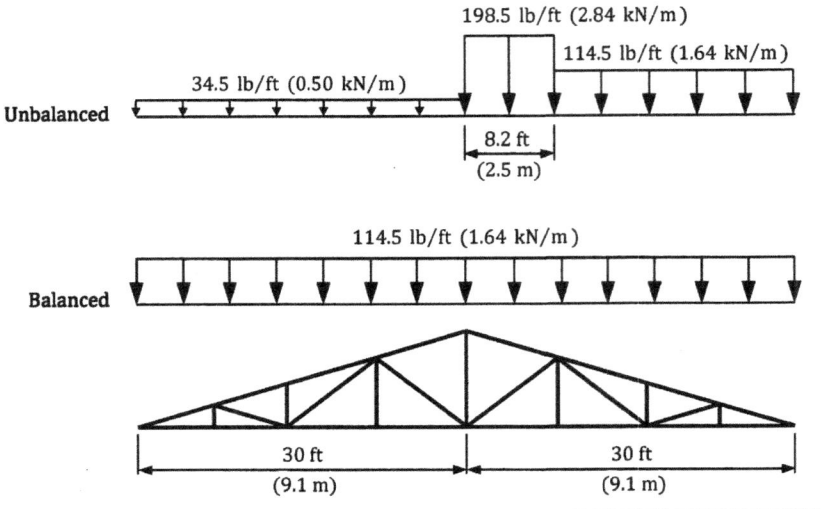

Figure 3.29 Balanced and unbalanced snow loads on a typical roof truss in Example 3.5.

Step 14—Consider snow loads on open-frame equipment structures Sec. 3.17

Not applicable.

The balanced and unbalanced snow loads on a typical roof truss are depicted in Fig. 3.29.

3.18.6 Example 3.6—Calculation of Design Snow Loads for a Curved Roof

Determine the design snow loads for the commercial building in Fig. 3.30 given the design data in Table 3.11.

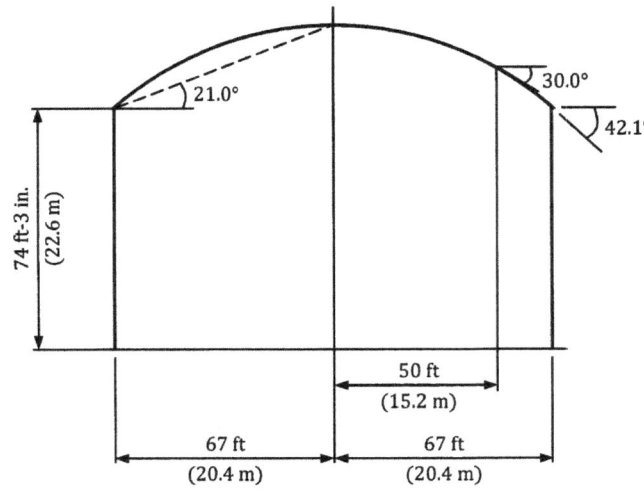

Figure 3.30 Elevation of the building with a curved roof in Example 3.6.

Location	Los Alamos, NM
Surface roughness	B
Occupancy	Commercial
Thermal condition	Unheated
Roof exposure	Partially exposed
Roof surface	Smooth rubber membrane
Roof obstructions	None
Roof framing	All structural members are simply supported

TABLE 3.11 Design Data for Example 3.6

Solution

Step 1—Determine the ground snow load, p_g   ASCE/SEI Table 7.2-6

The ground snow load is equal to 30 lb/ft² (1.44 kN/m²).

Step 2—Determine the flat roof snow load, p_f   Fig. 3.2

- Step 2a—Determine the surface roughness category

  From the design data, the surface roughness category is given as B.

- Step 2b—Determine the exposure of the roof

  From the design data, the roof exposure is given as partially exposed.

- Step 2c—Determine the exposure factor, C_e

  Given a surface roughness category of B and a partially exposed roof exposure, $C_e = 1.0$ from ASCE/SEI Table 7.3-1.

- Step 2d—Determine the thermal factor, C_t

  From the design data, the building is unheated. Therefore, $C_t = 1.2$ from ASCE/SEI Table 7.3-2.

- Step 2e—Determine the Risk Category of the building

  From the design data, the building has a commercial occupancy. Therefore, the Risk Category is II from ASCE/SEI Table 1.5-1.

- Step 2f—Determine the importance factor, I_s

  For a Risk Category II building, $I_s = 1.0$ from ASCE/SEI Table 1.5-2.

- Step 2g—Determine the flat roof snow load, p_f

$$p_f = 0.7 C_e C_t I_s p_g = 0.7 \times 1.0 \times 1.2 \times 1.0 \times 30.0 = 25.2 \text{ lb/ft}^2 \qquad (3.1)$$

$$p_f = 0.7 C_e C_t I_s p_g = 0.7 \times 1.0 \times 1.2 \times 1.0 \times 1.44 = 1.21 \text{ kN/m}^2$$

Step 3—Determine the minimum snow load for low-slope roofs, p_m   Sec. 3.6

A minimum snow load applies to curved roofs where the vertical angle from the eaves to the crown is less than 10 degrees (see Fig. 3.3). Because the vertical angle in this example is equal to 21 degrees, minimum roof snow loads do not apply.

Step 4—Determine the sloped roof (balanced) snow load, p_s   Sec. 3.7

- Step 4a—Determine the slope factor, C_s Fig. 3.5

It is determined in Step 2d that $C_t = 1.2$ (cold roof).

From the design data, there are no obstructions inhibiting the snow from sliding off the roof. Also, because the roof is cold, ice dams need not be considered (if an ice dam can form on a warm roof, it is considered to be an obstruction).

From the design data, the roof surface is a smooth rubber membrane. According to ASCE/SEI 7.4, smooth rubber membranes are considered to be slippery.

Because the roof is unobstructed and slippery, use the dashed line in ASCE/SEI Figure 7.4-1c to determine C_s.

For a tangent slope of 42.1 degrees at the eaves (see Fig. 3.5 for the following equation):

$$C_{s|eave} = 1.0 - \frac{\text{slope} - 15°}{55°} = 1.0 - \frac{42.1° - 15°}{55°} = 0.51$$

At a tangent slope of 30 degrees, which is located at a distance of 50 ft (15.2 m) from the crown:

$$C_{s|30} = 1.0 - \frac{\text{slope} - 15°}{55°} = 1.0 - \frac{30° - 15°}{55°} = 0.73$$

Slope factor $C_s = 1.0$ where the tangent slope is less than or equal to 15 degrees. This occurs at a distance of approximately 25.9 ft (7.9 m) from the crown. Therefore, $C_s = 1.0$ over a distance of $2 \times 25.9 = 51.8$ ft ($2 \times 7.9 = 15.8$ m) centered on the crown.

- Step 4b—Determine the sloped roof (balanced) snow load, p_s Fig. 3.10

Because the tangent slope at the eaves is equal to 42.1 degrees, use case 2 in ASCE/SEI Figure 7.4-2.

At the eaves:

$$p_s = C_{s|eave}p_f = 0.51 \times 25.2 = 12.9 \text{ lb/ft}^2$$
$$p_s = C_{s|eave}p_f = 0.51 \times 1.21 = 0.62 \text{ kN/m}^2$$

At the tangent slope of 30 degrees:

$$p_s = C_{s|30}p_f = 0.73 \times 25.2 = 18.4 \text{ lb/ft}^2$$
$$p_s = C_{s|30}p_f = 0.73 \times 1.21 = 0.88 \text{ kN/m}^2$$

Within the 51.8-ft (15.8-m) center portion of the roof, where $C_s = 1.0$:

$$p_s = C_s p_f = 1.0 \times 25.2 = 25.2 \text{ lb/ft}^2$$
$$p_s = C_s p_f = 1.0 \times 1.21 = 1.21 \text{ kN/m}^2$$

Step 5—Consider loads due to ice dams   Sec. 3.8

Because the roof has been determined to be a cold roof, ice dams and the accompanying uniform load need not be considered (ASCE/SEI 7.4.5).

Step 6—Consider partial loading Sec. 3.9

From the design data, the roof structural members are simply supported, so partial loading need not be considered for these members (ASCE/SEI 7.5).

Step 7—Consider unbalanced snow loads Sec. 3.10

Because the slope of the straight line from the eaves to the crown is greater than 10 degrees and less than 60 degrees, unbalanced snow loads must be considered (ASCE/SEI 7.6.2).

Unbalanced snow loads for this roof are given in case 2 of ASCE/SEI Figure 7.4-2 because the tangent slope at the eave is between 30 and 70 degrees (see Fig. 3.10).

- Windward side

  The unbalanced load is equal to zero.

- Leeward side

  At the eaves: $2p_f(C_{s|eave}/C_e) = 2 \times 25.2 \times (0.51/1.0) = 25.7$ lb/ft^2 [in S.I.: $2 \times 1.21 \times (0.51/1.0) = 1.23$ kN/m^2].

  At the tangent slope of 30 degrees:

  $2p_f(C_{s|30}/C_e) = 2 \times 25.2 \times (0.73/1.0) = 36.8$ lb/ft^2 [in S.I.: $2 \times 1.21 \times (0.73/1.0) = 1.77$ kN/m^2].

  At the crown: $0.5p_f = 0.5 \times 25.2 = 12.6$ lb/ft^2 [in S.I.: $0.5 \times 1.21 = 0.61$ kN/m^2].

Step 8—Consider drifts on lower roofs Sec. 3.11

Not applicable.

Step 9—Consider drifts on roof projections and parapets Sec. 3.12

Not applicable.

Step 10—Consider sliding snow loads Sec. 3.13

Not applicable.

Step 11—Consider rain-on-snow surcharge loads Sec. 3.14

A rain-on-snow surcharge load of 5 lb/ft^2 (0.24 kN/m^2) is required for locations where (1) the ground snow load, p_g, is 20 lb/ft^2 (0.96 kN/m^2) or less and greater than zero and (2) the roof slope is less than $W/50$ (ASCE/SEI 7.10).

In this example, $p_g = 30$ lb/ft^2 (1.44 kN/m^2), which is greater than 20 lb/ft^2 (0.96 kN/m^2), so a rain-on-snow load is not required.

Step 12—Consider ponding instability Sec. 3.15

Because the roof slope is greater than 1 inch per foot (4.76 degrees) at all locations, the roof is not susceptible to ponding, and progressive roof deflection and ponding instability from snow meltwater need not be investigated (see ASCE/SEI 7.11 and 8.4).

Step 13—Consider snow loads on existing roofs Sec. 3.16

Not applicable.

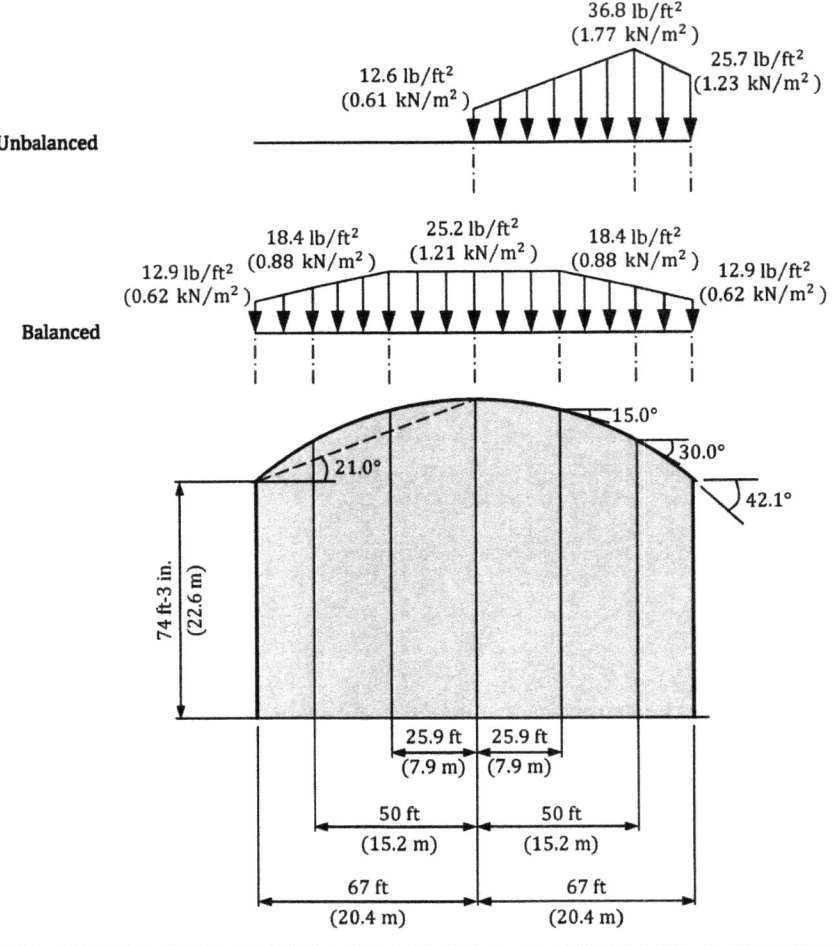

FIGURE 3.31 Balanced and unbalanced snow loads for the building in Example 3.6.

Step 14—Consider snow loads on open-frame equipment structures   Sec. 3.17

Not applicable.

The balanced and unbalanced snow loads on the roof of the building in this example are depicted in Fig. 3.31.

3.18.7 Example 3.7—Calculation of Design Snow Loads for a Sawtooth Roof

Determine the design snow loads for the industrial building in Fig. 3.32 given the design data in Table 3.12.

Solution

Step 1—Determine the ground snow load, p_g   ASCE/SEI Table 7.2-1

The ground snow load is equal to 50 lb/ft² (2.39 kN/m²).

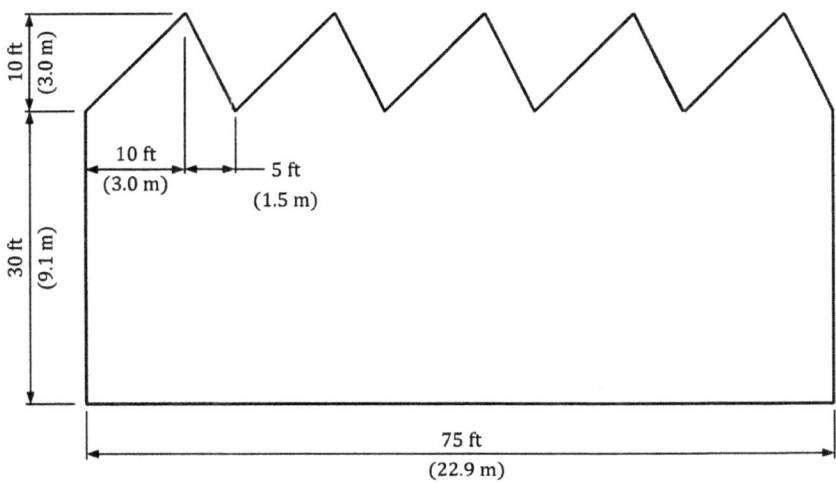

FIGURE 3.32 Elevation of the building with a sawtooth roof in Example 3.7.

Location	Anchorage, AK
Surface roughness	D
Occupancy	Industrial facility manufacturing hazardous chemicals
Thermal condition	Kept just above freezing
Roof exposure	Partially exposed
Roof surface	Glass
Roof obstructions	None
Roof framing	All structural members are simply supported

TABLE 3.12 Design Data for Example 3.7

Step 2—Determine the flat roof snow load, p_f   Fig. 3.2

- Step 2a—Determine the surface roughness category

  From the design data, the surface roughness category is given as D.

- Step 2b—Determine the exposure of the roof

  From the design data, the roof exposure is given as partially exposed.

- Step 2c—Determine the exposure factor, C_e

  Given a surface roughness category of D and a partially exposed roof exposure, $C_e = 0.9$ from ASCE/SEI Table 7.3-1.

- Step 2d—Determine the thermal factor, C_t

  From the design data, the building is kept just above freezing. Therefore, $C_t = 1.1$ from ASCE/SEI Table 7.3-2.

- Step 2e—Determine the Risk Category of the building

  From the design data, the building has an industrial occupancy where hazardous chemicals are manufactured. The authority having jurisdiction has determined the Risk Category to be III from ASCE/SEI Table 1.5-1.

- Step 2f—Determine the importance factor, I_s

  For a Risk Category III building, $I_s = 1.1$ from ASCE/SEI Table 1.5-2.

- Step 2g—Determine the flat roof snow load, p_f

$$p_f = 0.7C_eC_tI_sp_g = 0.7 \times 0.9 \times 1.1 \times 1.1 \times 50.0 = 38.1 \text{ lb/ft}^2 \tag{3.1}$$

$$p_f = 0.7C_eC_tI_sp_g = 0.7 \times 0.9 \times 1.1 \times 1.1 \times 2.39 = 1.82 \text{ kN/m}^2$$

Step 3—Determine the minimum snow load for low-slope roofs, p_m   Sec. 3.6

   Not applicable for sawtooth roofs.

Step 4—Determine the sloped roof (balanced) snow load, p_s   Sec. 3.7

- Step 4a—Determine the slope factor, C_s Fig. 3.5

  In accordance with ASCE/SEI 7.4.4, $C_s = 1.0$ for sawtooth roofs.

- Step 4b—Determine the sloped roof (balanced) snow load, p_s Fig. 3.11

$$p_s = C_sp_f = 1.0 \times 38.1 = 38.1 \text{ lb/ft}^2 \tag{3.2}$$

$$p_s = C_sp_f = 1.0 \times 1.82 = 1.82 \text{ kN/m}^2$$

Step 5—Consider loads due to ice dams   Sec. 3.8

   Not applicable.

Step 6—Consider partial loading   Sec. 3.9

   From the design data, the roof structural members are simply supported, so partial loading need not be considered for these members (ASCE/SEI 7.5).

Step 7—Consider unbalanced snow loads   Sec. 3.10

   The roof slopes are equal to 45 degrees and 63.4 degrees. Because these slopes are greater than 1.79 degrees, unbalanced snow loads must be considered (ASCE/SEI 7.6.3).

   Unbalanced snow loads for this roof are given in ASCE/SEI Figure 7.6-3 (see Fig. 3.11).

- At the ridge:

$$0.5p_f = 0.5 \times 38.1 = 19.1 \text{ lb/ft}^2$$

$$0.5p_f = 0.5 \times 1.82 = 0.91 \text{ kN/m}^2$$

- At the valley:

$$2p_f/C_e = (2 \times 38.1)/0.9 = 84.7 \text{ lb/ft}^2$$

The load at the valley is limited by the space available for snow accumulation.

$$\gamma = 0.13p_g + 14 = (0.13 \times 50.0) + 14 = 20.5 \ \text{lb/ft}^3 < 30.0 \ \text{lb/ft}^3 \tag{3.5}$$

Ridge height $h_r = 10.0$ ft
    Maximum load at the valley:

$$0.5p_f + \gamma h_r = (0.5 \times 38.1) + (20.5 \times 10.0) = 224.1 \ \text{lb/ft}^2$$

Because the unbalanced load of 84.7 lb/ft² at the valley is less than 224.1 lb/ft², use 84.7 lb/ft² as the load at the valley.
    In S.I.:

$$2p_f/C_e = (2 \times 1.82)/0.9 = 4.04 \ \text{kN/m}^2$$

The load at the valley is limited by the space available for snow accumulation.

$$\gamma = 0.426p_g + 2.2 = (0.426 \times 2.39) + 2.2 = 3.22 \ \text{kN/m}^3 < 4.70 \ \text{kN/m}^3 \tag{3.6}$$

Ridge height $h_r = 3.0$ m
    Maximum load at the valley:

$$0.5p_f + \gamma h_r = (0.5 \times 1.82) + (3.22 \times 3.0) = 10.6 \ \text{kN/m}^2$$

Because the unbalanced load of 4.04 kN/m² at the valley is less than 10.6 kN/m², use 4.04 kN/m² as the load at the valley.

Step 8—Consider drifts on lower roofs   Sec. 3.11
    Not applicable.

Step 9—Consider drifts on roof projections and parapets   Sec. 3.12
    Not applicable.

Step 10—Consider sliding snow loads   Sec. 3.13
    Not applicable.

Step 11—Consider rain-on-snow surcharge loads   Sec. 3.14

    A rain-on-snow surcharge load of 5 lb/ft² (0.24 kN/m²) is required for locations where (1) the ground snow load, p_g, is 20 lb/ft² (0.96 kN/m²) or less and greater than zero and (2) the roof slope is less than $W/50$ (ASCE/SEI 7.10).
    In this example, $p_g = 50$ lb/ft² (2.39 kN/m²), which is greater than 20 lb/ft² (0.96 kN/m²), so a rain-on-snow load is not required.

Step 12—Consider ponding instability   Sec. 3.15

    Because the roof slope is greater than 1 inch per foot (4.76 degrees) at all locations, the roof is not susceptible to ponding, and progressive roof deflection and ponding instability from snow meltwater need not be investigated (see ASCE/SEI 7.11 and 8.4).

Step 13—Consider snow loads on existing roofs   Sec. 3.16
    Not applicable.

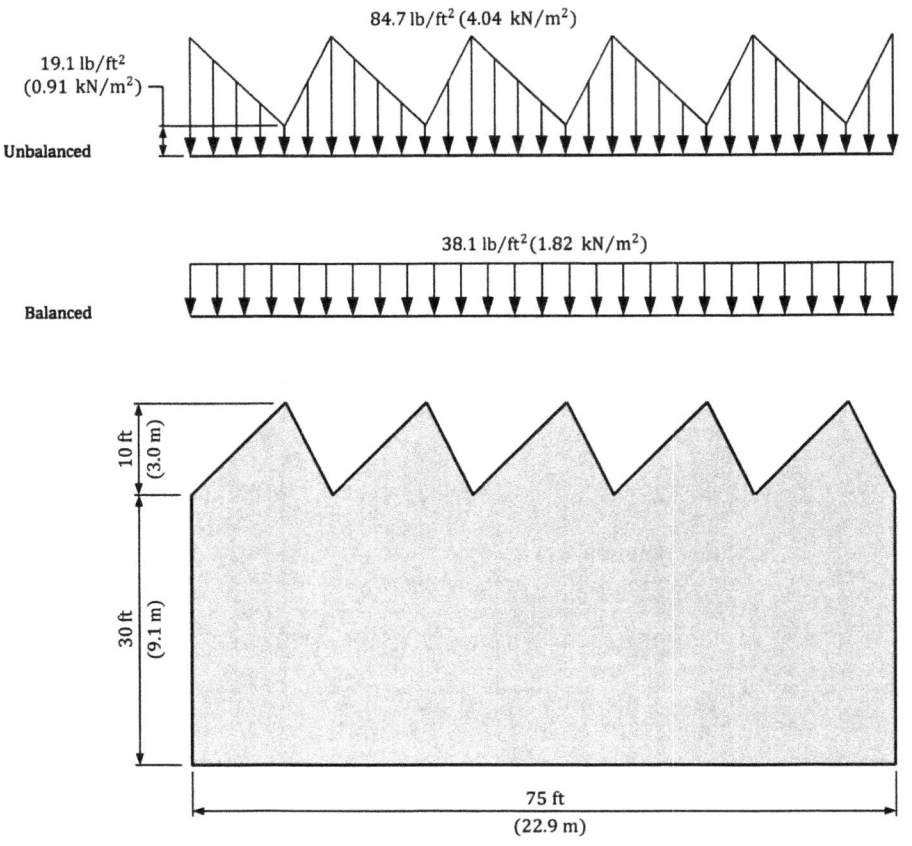

FIGURE 3.33 Balanced and unbalanced snow loads for the building in Example 3.7.

Step 14—Consider snow loads on open-frame equipment structures   Sec. 3.17

Not applicable.

The balanced and unbalanced snow loads on the roof of the building in this example are depicted in Fig. 3.33.

3.18.8   Example 3.8—Calculation of Design Snow Loads for an Upper Roof and a Lower Roof Including Sliding Snow

Determine the design snow loads on the upper roof and the lower roof of the residential building in Fig. 3.34 given the design data in Table 3.13.

Solution

Snow Loads on Upper Roof

Step 1—Determine the ground snow load, p_g   ASCE/SEI Table 7.2-8

The ground snow load is equal to 70 lb/ft² (3.35 kN/m²).

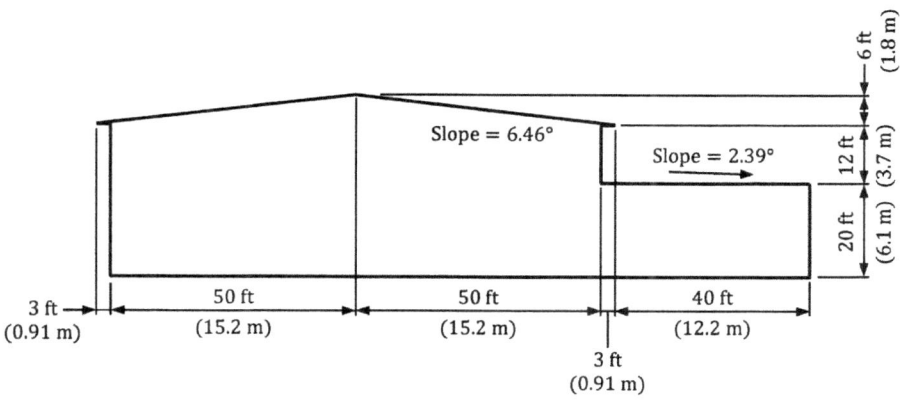

Figure 3.34 Elevation of the residential building in Example 3.8.

Location	Amherst, NH
Surface roughness	C
Occupancy	Residential
Thermal condition	Cold, ventilated roof with $R > 25$ ft²h°F/Btu (4.4 m²K/W)
Roof exposure	Sheltered
Roof surface	Metal
Roof obstructions	None
Roof framing	For both roofs, simply supported primary and secondary members with the secondary members spanning parallel to the free-draining edges of the roof

Table 3.13 Design Data for Example 3.8

Step 2—Determine the flat roof snow load, p_f   Fig. 3.2

- Step 2a—Determine the surface roughness category

  From the design data, the surface roughness category is given as C.

- Step 2b—Determine the exposure of the roof

  From the design data, the roof exposure is given as sheltered.

- Step 2c—Determine the exposure factor, C_e

  Given a surface roughness category of C and a sheltered roof exposure, $C_e = 1.1$ from ASCE/SEI Table 7.3-1.

- Step 2d—Determine the thermal factor, C_t

  From the design data, the building has a cold, ventilated roof with $R > 25$ ft²h°F/Btu (4.4 m²K/W). Therefore, $C_t = 1.1$ from ASCE/SEI Table 7.3-2.

- Step 2e—Determine the Risk Category of the building

  From the design data, the building has a residential occupancy. Therefore, the Risk Category is II from ASCE/SEI Table 1.5-1.

- Step 2f—Determine the importance factor, I_s

  For a Risk Category II building, $I_s = 1.0$ from ASCE/SEI Table 1.5-2.

- Step 2g—Determine the flat roof snow load, p_f

$$p_f = 0.7C_eC_tI_sp_g = 0.7 \times 1.1 \times 1.1 \times 1.0 \times 70.0 = 59.3 \text{ lb/ft}^2 \qquad (3.1)$$

$$p_f = 0.7C_eC_tI_sp_g = 0.7 \times 1.1 \times 1.1 \times 1.0 \times 3.35 = 2.84 \text{ kN/m}^2$$

Step 3—Determine the minimum snow load for low-slope roofs, p_m   Sec. 3.6

A minimum snow load applies to gable roofs with slopes less than 15 degrees (see Fig. 3.3). Because the upper roof slope in this example is equal to 6.46 degrees, minimum roof snow loads must be considered:

For $p_g = 70.0 \text{ lb/ft}^2 > 20.0 \text{ lb/ft}^2$, $p_m = 20I_s = 20.0 \times 1.0 = 20.0 \text{ lb/ft}^2$ Table 3.3

For $p_g = 3.35 \text{ kN/m}^2 > 0.96 \text{ kN/m}^2$, $p_m = 0.96I_s = 0.96 \times 1.0 = 0.96 \text{ kN/m}^2$

Step 4—Determine the sloped roof (balanced) snow load, p_s   Sec. 3.7

- Step 4a—Determine the slope factor, C_s Fig. 3.5

  It is determined in Step 2d that $C_t = 1.1$ (cold roof).

  From the design data, there are no obstructions inhibiting the snow from sliding off the roof. Also, because the roof is cold, ice dams need not be considered (if an ice dam can form on a warm roof, it is considered to be an obstruction).

  From the design data, the roof surface is metal. According to ASCE/SEI 7.4, metal roof surfaces are considered to be slippery surfaces.

  Because the roof is unobstructed and slippery, use the dashed line in ASCE/SEI Figure 7.4-1b to determine C_s:

  For a roof slope of 6.46 degrees, which is less than 10 degrees, $C_s = 1.0$.

- Step 4b—Determine the sloped roof (balanced) snow load, p_s Fig. 3.9

$$p_s = C_sp_f = 1.0 \times 59.3 = 59.3 \text{ lb/ft}^2 \qquad (3.2)$$

$$p_s = C_sp_f = 1.0 \times 2.84 = 2.84 \text{ kN/m}^2$$

Step 5—Consider loads due to ice dams   Sec. 3.8

Because the roof has been determined to be a cold roof, ice dams and the accompanying uniform load need not be considered (ASCE/SEI 7.4.5).

Step 6—Consider partial loading   Sec. 3.9

From the design data, the primary and secondary framing members are simply supported, so partial loading need not be considered for these members (ASCE/SEI 7.5).

Step 7—Consider unbalanced snow loads Sec. 3.10

Unbalanced snow loads need not be considered for gable roofs with a slope exceeding 30.3 degrees or with a slope less than 2.39 degrees (ASCE/SEI 7.6.1). The upper roof slope in this example is 6.46 degrees, which means unbalanced loads must be considered.

Because $W = 53$ ft (16.2 m) > 20 ft (6.1 m), the unbalanced load consists of the following (see Fig. 3.9):

- Windward side

    Unbalanced load $= 0.3p_s = 0.3 \times 59.3 = 17.8$ lb/ft² is applied over the entire length of the windward side.

- Leeward side

$$h_d = \sqrt{I_s}\left\{[0.43(W)^{1/3}(p_g+10)^{1/4}]-1.5\right\} \qquad \text{Eq. (3.3)}$$

$$= \sqrt{1.0} \times \left\{[0.43 \times (53.0)^{1/3} \times (70.0+10)^{1/4}]-1.5\right\} = 3.3 \text{ ft}$$

$$\gamma = 0.13p_g + 14 = (0.13 \times 70.0) + 14 = 23.1 \text{ lb/ft}^3 < 30.0 \text{ lb/ft}^3 \qquad (3.5)$$

$$S = \text{roof slope run for a rise of one} = 1/\tan 6.46° = 8.8$$

Balanced load $= p_s = 59.3$ lb/ft² is applied over the entire length of the leeward side.

Uniform pressure of $h_d\gamma/\sqrt{S} = (3.3 \times 23.1)/\sqrt{8.8} = 25.7$ lb/ft² is applied from the ridge a distance of $8h_d\sqrt{S}/3 = (8 \times 3.3 \times \sqrt{8.8})/3 = 26.1$ ft.

In S.I.:

- Windward side

    Unbalanced load $= 0.3p_s = 0.3 \times 2.84 = 0.85$ kN/m² is applied over the entire length of the windward side.

- Leeward side

$$h_d = \sqrt{I_s}\left\{[0.42(W)^{1/3}(p_g+0.479)^{1/4}]-0.457\right\} \qquad (3.4)$$

$$= \sqrt{1.0} \times \left\{[0.42 \times (16.2)^{1/3} \times (3.35+0.479)^{1/4}]-0.457\right\} = 1.0 \text{ m}$$

$$\gamma = 0.426p_g + 2.2 = (0.426 \times 3.35) + 2.2 = 3.6 \text{ kN/m}^3 < 4.7 \text{ kN/m}^3 \qquad (3.6)$$

$$S = \text{roof slope run for a rise of one} = 1/\tan 6.46° = 8.8$$

Balanced load $= p_s = 2.84$ kN/m² is applied over the entire length of the leeward side.

Uniform pressure of $h_d\gamma/\sqrt{S} = (1.0 \times 3.6)/\sqrt{8.8} = 1.21$ kN/m² is applied from the ridge a distance of $8h_d\sqrt{S}/3 = (8 \times 1.0 \times \sqrt{8.8})/3 = 7.9$ m.

Step 8—Consider drifts on lower roofs Sec. 3.11

See the section of this solution below for snow loads on the lower roof.

Step 9—Consider drifts on roof projections and parapets   Sec. 3.12

Not applicable.

Step 10—Consider sliding snow loads   Sec. 3.13

See the section of this solution below for snow loads on the lower roof.

Step 11—Consider rain-on-snow surcharge loads   Sec. 3.14

A rain-on-snow surcharge load of 5 lb/ft² (0.24 kN/m²) is required for locations where (1) the ground snow load, p_g, is 20 lb/ft² (0.96 kN/m²) or less and greater than zero and (2) the roof slope is less than $W/50$ (ASCE/SEI 7.10).

In this example, p_g = 70 lb/ft² (3.35 kN/m²), which is greater than 20 lb/ft² (0.96 kN/m²), so a rain-on-snow load is not required.

Step 12—Consider ponding instability   Sec. 3.15

Because the secondary members are parallel to the free-draining edges of the roof and the roof slope is greater than 1 inch per foot (4.76 degrees), the bay is not susceptible to ponding, and progressive roof deflection and ponding instability from snow meltwater need not be investigated (see Fig. 2.8).

Step 13—Consider snow loads on existing roofs   Sec. 3.16

Not applicable.

Step 14—Consider snow loads on open-frame equipment structures   Sec. 3.17

Not applicable.

The balanced and unbalanced snow loads on the upper roof of the building in this example are depicted in Fig. 3.35.

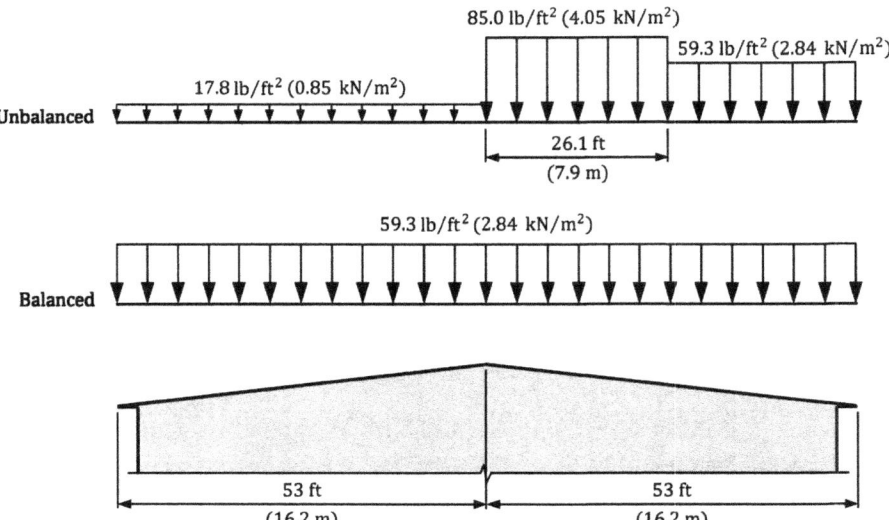

FIGURE 3.35 Balanced and unbalanced snow loads for the upper roof in Example 3.8.

Snow Loads on Lower Roof

Steps 1 through 6—Results are the same as those for the upper.

$$p_g = 70 \text{ lb/ft}^2 \ (3.35 \text{ kN/m}^2)$$

$$C_e = 1.1$$

$$C_t = 1.1$$

$$I_s = 1.0$$

$$p_f = p_s = 59.3 \text{ lb/ft}^2 \ (2.84 \text{ kN/m}^2)$$

$$p_m = 20.0 \text{ lb/ft}^2 \ (0.96 \text{ kN/m}^2)$$

$$\gamma = 23.1 \text{ lb/ft}^3 (3.6 \text{ kN/m}^3)$$

Ice dam loads and partial loading need not be considered.

Step 7—Consider unbalanced snow loads   Sec. 3.10

Because this is a monoslope roof, unbalanced snow loads need not be considered.

Step 8—Consider drifts on lower roofs   Sec. 3.11

Use Fig. 3.15 to determine the leeward and windward drifts forming on the lower roof.

• Step 8a—Determine h_b

$$h_b = p_s/\gamma = 59.3/23.1 = 2.6 \text{ ft}$$

• Step 8b—Determine h_c

$$h_c = h_{step} - h_b = [12.0 - (43.0 \times \tan 2.39°)] - 2.6 = 7.6 \text{ ft}$$

• Step 8c—Determine if drift loads must be considered

Drift loads must be considered where $h_c/h_b \geq 0.2$.

$$h_c/h_b = 7.6/2.6 = 2.9 > 0.2$$

Therefore, drift loads must be considered.

• Step 8d—Determine the leeward drift height, $h_{d,leeward}$

Use the following equation, which is applicable where $\ell_{upper} \geq 20$ ft:

$$h_{d,leeward} = \sqrt{I_s} \left\{ [0.43(\ell_{upper})^{1/3}(p_g + 10)^{1/4}] - 1.5 \right\} \leq 0.6\ell_{lower} \qquad \text{Table 3.5}$$

$$= \sqrt{1.0} \times \left\{ [0.43 \times (106.0)^{1/3} \times (70.0 + 10)^{1/4}] - 1.5 \right\}$$

$$= 4.6 \text{ ft} < 0.6 \times 43.0 = 25.8 \text{ ft}$$

- Step 8e—Determine the windward drift height, $h_{d,windward}$
  Use the following equation, which is applicable where $\ell_{lower} \geq 20$ ft:

$$h_{d,windward} = 0.75\sqrt{I_s}\left\{[0.43(\ell_{lower})^{1/3}(p_g+10)^{1/4}]-1.5\right\}$$ Table 3.5

$$= 0.75 \times \sqrt{1.0} \times \left\{[0.43 \times (43.0)^{1/3} \times (70.0+10)^{1/4}]-1.5\right\} = 2.3 \text{ ft}$$

- Step 8f—Determine h_d

$$h_d = \text{larger of } (h_{d,leeward}, h_{d,windward}) = 4.6 \text{ ft}$$

- Step 8g—Determine the drift width, w

Because $h_d = 4.6 \text{ ft} < h_c = 7.6 \text{ ft}$, $w = 4h_d = 18.4 \text{ ft}$

- Step 8h—Determine the drift load, p_d

$$p_d = \gamma h_d = 23.1 \times 4.6 = 106.3 \text{ lb/ft}^2$$ (3.8)

- Step 8i—Determine the total snow load at the step

$$p_{total} = p_s + p_d = 59.3 + 106.3 = 165.6 \text{ lb/ft}^2$$

In S.I.:

- Step 8a—Determine h_b

$$h_b = p_s/\gamma = 2.84/3.6 = 0.8 \text{ m}$$

- Step 8b—Determine h_c

$$h_c = h_{step} - h_b = [3.7 - (13.1 \times \tan 2.39°)] - 0.8 = 2.4 \text{ m}$$

- Step 8c—Determine if drift loads must be considered
  Drift loads must be considered where $h_c/h_b \geq 0.2$.

$$h_c/h_b = 2.4/0.8 = 3.0 > 0.2$$

Therefore, drift loads must be considered.

- Step 8d—Determine the leeward drift height, $h_{d,leeward}$
  Use the following equation, which is applicable where $\ell_{upper} \geq 6.1$ m:

$$h_{d,leeward} = \sqrt{I_s}\left\{[0.42(\ell_{upper})^{1/3}(p_g+0.479)^{1/4}]-0.457\right\} \leq 0.6\ell_{lower}$$ Table 3.5

$$= \sqrt{1.0} \times \left\{[0.42 \times (32.3)^{1/3} \times (3.35+0.479)^{1/4}]-0.457\right\}$$

$$= 1.4 \text{ m} < 0.6 \times 13.1 = 7.9 \text{ m}$$

- Step 8e—Determine the windward drift height, $h_{d,windward}$

  Use the following equation, which is applicable where $\ell_{lower} \geq 6.1$ m:

  $$h_{d,windward} = 0.75\sqrt{I_s}\left\{[0.42(\ell_{lower})^{1/3}(p_g + 0.479)^{1/4}] - 0.457\right\} \qquad \text{Table 3.5}$$

  $$= 0.75 \times \sqrt{1.0} \times \left\{[0.42 \times (13.1)^{1/3} \times (3.35 + 0.479)^{1/4}] - 0.457\right\} = 0.7 \text{ m}$$

- Step 8f—Determine h_d

  $$h_d = \text{larger of } (h_{d,leeward},\ h_{d,windward}) = 1.4 \text{ m}$$

- Step 8g—Determine the drift width, w

  Because $h_d = 1.4$ m $< h_c = 2.4$ m, $w = 4h_d = 5.6$ m

- Step 8h—Determine the drift load, p_d

  $$p_d = \gamma h_d = 3.6 \times 1.4 = 5.04 \text{ kN/m}^2 \qquad (3.8)$$

- Step 8i—Determine the total snow load at the step

  $$p_{total} = p_s + p_d = 2.84 + 5.04 = 7.88 \text{ kN/m}^2$$

Step 9—Consider drifts on roof projections and parapets   Sec. 3.12

Not applicable.

Step 10—Consider sliding snow loads   Sec. 3.13

The provisions of ASCE/SEI 7.9 are used to determine if a load due to snow sliding off the upper roof on to the lower roof must be considered.

Loads caused by snow sliding must be considered because the upper roof is slippery with a slope greater than ¼ on 12 (1.19 degrees).

The sliding snow load is equal to the following where $\ell_{lower} > 15$ ft (4.6 m) (see Fig. 3.19):

$$p_{sliding} = 0.4p_f W/15 = (0.4 \times 59.3 \times 53.0)/15 = 83.8 \text{ lb/ft}^2$$

$$p_{sliding} = 0.4p_f W/4.6 = (0.4 \times 2.84 \times 16.1)/4.6 = 3.98 \text{ kN/m}^2$$

The total snow load over the 15-ft (4.6-m) width is equal to the sliding snow load plus the balanced snow load:

$$p_{total} = p_s + (0.4p_f W/15) = 59.3 + 83.8 = 143.1 \text{ lb/ft}^2 \qquad (3.9)$$

$$p_{total} = p_s + (0.4p_f W/4.6) = 2.84 + 3.98 = 6.82 \text{ kN/m}^2 \qquad (3.10)$$

The depth of snow for the total snow load $= p_{total}/\gamma = 143.1/23.1 = 6.2$ ft (in S.I.: $p_{total}/\gamma = 6.82/3.6 = 1.9$ m). This depth is less than the distance from the eave of the upper roof to the top of the lower roof at the interface, which is equal to $h_{step} = 12.0 - (43.0 \times \tan 2.39°) = 10.2$ ft [in S.I.: $h_{step} = 3.7 - (13.1 \times \tan 2.39°) = 3.2$ m]. Therefore, sliding snow is not blocked and the full load can be developed over the 15-ft (4.6-m) width.

Step 11—Consider rain-on-snow surcharge loads   Sec. 3.14

A rain-on-snow surcharge load of 5 lb/ft² (0.24 kN/m²) is required for locations where (1) the ground snow load, p_g, is 20 lb/ft² (0.96 kN/m²) or less and greater than zero and (2) the roof slope is less than $W/50$ (ASCE/SEI 7.10).

In this example, p_g = 70 lb/ft² (3.35 kN/m²), which is greater than 20 lb/ft² (0.96 kN/m²), so a rain-on-snow load is not required.

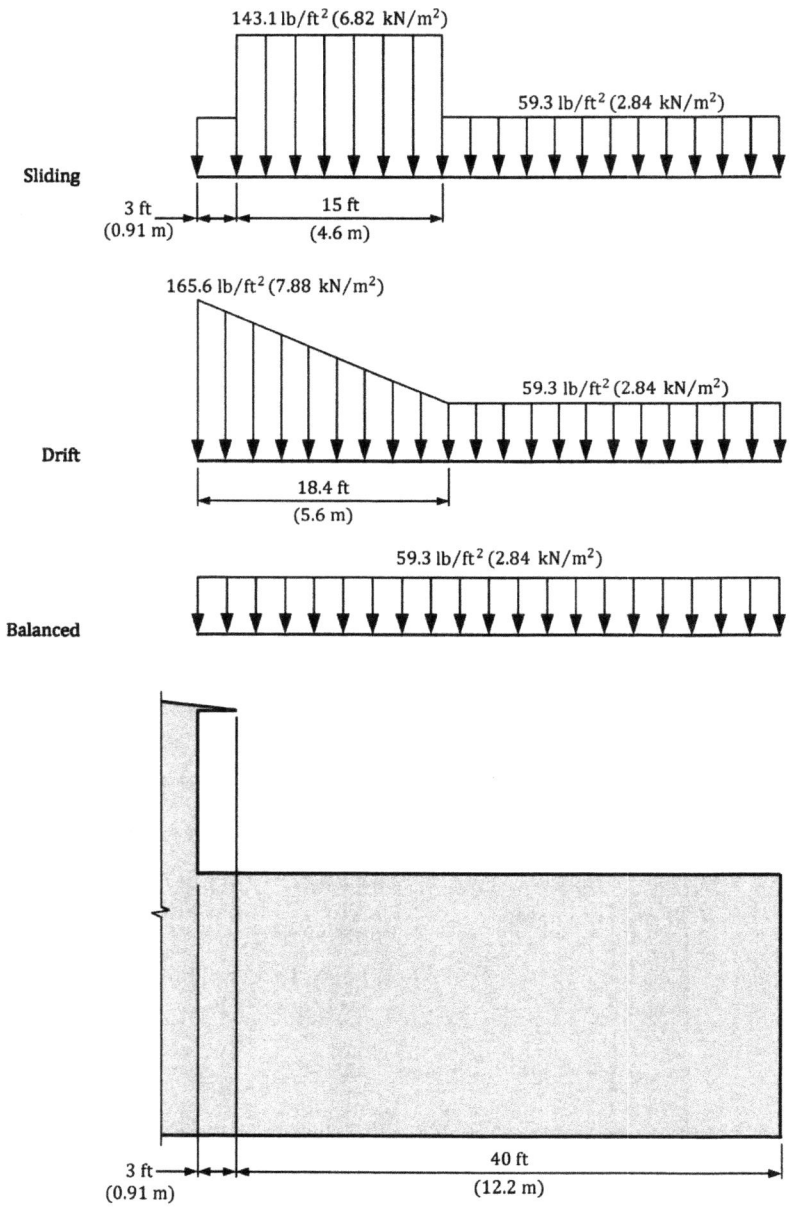

FIGURE 3.36   Balanced, drift, and sliding snow loads for the lower roof in Example 3.8.

Step 12—Consider ponding instability   Sec. 3.15

Because the secondary members are parallel to the free-draining edge of the roof and the roof slope is less than 1 inch per foot (4.76 degrees), the bay is susceptible to ponding, and progressive roof deflection and ponding instability from snow meltwater must be investigated (see Fig. 2.8).

Step 13—Consider snow loads on existing roofs   Sec. 3.16

Not applicable.

Step 14—Consider snow loads on open-frame equipment structures   Sec. 3.17

Not applicable.

The balanced, drift, and sliding snow loads on the lower roof of the building in this example are depicted in Fig. 3.36.

3.18.9 Example 3.9—Calculation of Design Snow Loads for a Lower Roof Including Sliding Snow and Horizontal Separation with an Adjacent Building

Determine the design snow loads on the lower roof of the residential building in Fig. 3.37. Use the design data in Table 3.13.

Solution

Steps 1 through 7—Results are the same as in Example 3.8 for the lower roof

$$p_g = 70 \text{ lb/ft}^2 \ (3.35 \text{ kN/m}^2)$$

$$C_e = 1.1$$

$$C_t = 1.1$$

$$I_s = 1.0$$

$$p_f = p_s = 59.3 \text{ lb/ft}^2 \ (2.84 \text{ kN/m}^2)$$

$$p_m = 20.0 \text{ lb/ft}^2 \ (0.96 \text{ kN/m}^2)$$

$$\gamma = 23.1 \text{ lb/ft}^3 (3.6 \text{ kN/m}^3)$$

Ice dam loads, partial loading, and unbalanced snow loads need not be considered.

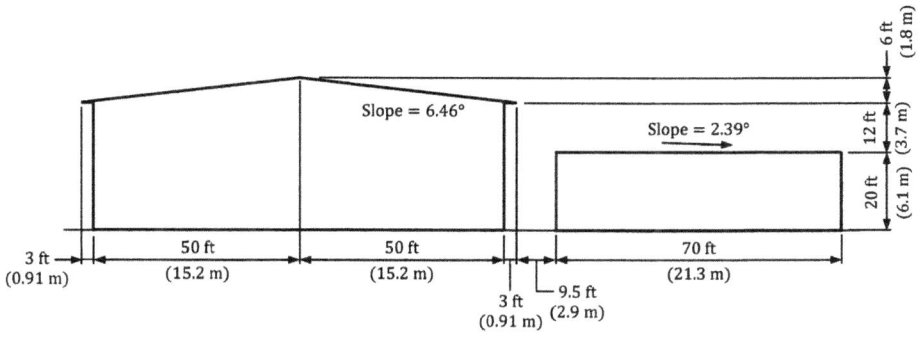

FIGURE 3.37 Elevation of the residential buildings in Example 3.9.

Step 8—Consider drifts on lower roofs   Sec. 3.11

Use Fig. 3.16 to determine the leeward and windward drifts forming on the lower roof.

- Step 8a—Determine if a leeward drift on the lower adjacent roof must be considered

  Leeward drifts must be considered where $s < 20$ ft and $s < 6h$.

  $$s = 9.5 \text{ ft} < 20 \text{ ft and } s < 6h = 6 \times 12 = 72 \text{ ft}$$

  Therefore, a leeward drift must be considered.

- Step 8b—Determine the leeward drift height, $h_{d,\text{leeward}}$

  Use the following equation, which is applicable where $\ell_{\text{upper}} \geq 20$ ft:

  $$h_{d,\text{leeward}} = \sqrt{I_s}\left\{[0.43(\ell_{\text{upper}})^{1/3}(p_g + 10)^{1/4}] - 1.5\right\}$$
  $$= \sqrt{1.0} \times \left\{[0.43 \times (106.0)^{1/3} \times (70.0 + 10)^{1/4}] - 1.5\right\} = 4.6 \text{ ft}$$

- Step 8c—Determine the leeward drift load, p_d

  $$p_d = \text{smaller of} \begin{cases} \gamma h_{d,\text{leeward}} = 23.1 \times 4.6 = 106.3 \text{ lb/ft}^2 \\[2mm] \gamma[(6h - s)/6] = 23.1 \times [(6 \times 12.0) - 9.5]/6 = 240.6 \text{ lb/ft}^2 \end{cases}$$

- Step 8d—Determine the leeward drift width, w

  $$w = \text{smaller of} \begin{cases} 6h_{d,\text{leeward}} = 6 \times 4.6 = 27.6 \text{ ft} \\[2mm] 6h - s = (6 \times 12.0) - 9.5 = 62.5 \text{ ft} \end{cases}$$

- Step 8e—Determine the total snow load at the lower roof edge based on a leeward drift

  $$p_{\text{total}} = p_s + p_d = 59.3 + 106.3 = 165.6 \text{ lb/ft}^2$$

- Step 8f—Determine the windward drift height, $h_{d,\text{windward}}$

  Use the following equation, which is applicable where $\ell_{\text{lower}} \geq 20$ ft:

  $$h_{d,\text{windward}} = 0.75\sqrt{I_s}\left\{[0.43(\ell_{\text{lower}})^{1/3}(p_g + 10)^{1/4}] - 1.5\right\}$$
  $$= 0.75 \times \sqrt{1.0} \times \left\{[0.43 \times (70.0)^{1/3} \times (70.0 + 10)^{1/4}] - 1.5\right\} = 2.9 \text{ ft}$$

- Step 8g—Determine the windward drift load, p_d

  $$p_d = \gamma h_{d,\text{windward}} = 23.1 \times 2.9 = 67.0 \text{ lb/ft}^2$$

- Step 8h—Determine the truncated windward drift load at the edge of the lower roof, $p_{d,\text{truncated}}$

  $$p_{d,\text{truncated}} = \left(1 - \frac{s}{4h_{d,\text{windward}}}\right)p_d = \left(1 - \frac{9.5}{4 \times 2.9}\right) \times 67.0 = 12.1 \text{ lb/ft}^2$$

- Step 8i—Determine the windward drift width, w

$$w = 4h_{d,windward} - s = (4 \times 2.9) - 9.5 = 2.1 \text{ ft}$$

- Step 8j—Determine the total snow load at the lower roof edge based on a windward drift

$$p_{total} = p_s + p_{d,truncated} = 59.3 + 12.1 = 71.4 \text{ lb/ft}^2$$

In S.I.:

Step 8—Consider drifts on lower roofs   Sec. 3.11

Use Fig. 3.16 to determine the leeward and windward drifts forming on the lower roof.

- Step 8a—Determine if a leeward drift on the lower adjacent roof must be considered

Leeward drifts must be considered where $s < 6.1$ m and $s < 6h$.

$$s = 2.9 \text{ m} < 6.1 \text{ m and} < 6h = 6 \times 3.7 = 22.2 \text{ m}$$

Therefore, a leeward drift must be considered.

- Step 8b—Determine the leeward drift height, $h_{d,leeward}$

Use the following equation, which is applicable where $\ell_{upper} \geq 6.1$ m:

$$h_{d,leeward} = \sqrt{I_s} \left\{ [0.42(\ell_{upper})^{1/3}(p_g + 0.479)^{1/4}] - 0.457 \right\}$$

$$= \sqrt{1.0} \times \left\{ [0.42 \times (32.3)^{1/3} \times (3.35 + 0.479)^{1/4}] - 0.457 \right\} = 1.4 \text{ m}$$

- Step 8c—Determine the leeward drift load, p_d

$$p_d = \text{smaller of} \begin{cases} \gamma h_{d,leeward} = 3.6 \times 1.4 = 5.04 \text{ kN/m}^2 \\ \gamma[(6h - s)/6] = 3.6 \times [(6 \times 3.7) - 2.9]/6 = 11.6 \text{ kN/m}^2 \end{cases}$$

- Step 8d—Determine the leeward drift width, w

$$w = \text{smaller of} \begin{cases} 6h_{d,leeward} = 6 \times 1.4 = 8.4 \text{ m} \\ 6h - s = (6 \times 3.7) - 2.9 = 19.3 \text{ m} \end{cases}$$

- Step 8e—Determine the total snow load at the lower roof edge based on a leeward drift

$$p_{total} = p_s + p_d = 2.84 + 5.04 = 7.88 \text{ kN/m}^2$$

- Step 8f—Determine the windward drift height, $h_{d,windward}$

Use the following equation, which is applicable where $\ell_{lower} \geq 6.1$ m:

$$h_{d,windward} = 0.75\sqrt{I_s} \left\{ [0.42(\ell_{lower})^{1/3}(p_g + 0.479)^{1/4}] - 0.457 \right\}$$

$$= 0.75 \times \sqrt{1.0} \times \left\{ [0.42 \times (21.3)^{1/3} \times (3.35 + 0.479)^{1/4}] - 0.457 \right\} = 0.88 \text{ m}$$

- Step 8g—Determine the windward drift load, p_d

$$p_d = \gamma h_{d,\text{windward}} = 3.6 \times 0.88 = 3.2 \text{ kN/m}^2$$

- Step 8h—Determine the truncated windward drift load at the edge of the lower roof, $p_{d,\text{truncated}}$

$$p_{d,\text{truncated}} = \left(1 - \frac{s}{4h_{d,\text{windward}}}\right)p_d = \left(1 - \frac{2.9}{4 \times 0.88}\right) \times 3.2 = 0.56 \text{ kN/m}^2$$

- Step 8i—Determine the windward drift width, w

$$w = 4h_{d,\text{windward}} - s = (4 \times 0.88) - 2.9 = 0.62 \text{ m}$$

- Step 8j—Determine the total snow load at the lower roof edge based on a windward drift

$$p_{\text{total}} = p_s + p_{d,\text{truncated}} = 2.84 + 0.56 = 3.40 \text{ kN/m}^2$$

Step 9—Consider drifts on roof projections and parapets　　　　　　　Sec. 3.12

Not applicable.

Step 10—Consider sliding snow loads　　　　　　　Sec. 3.13

The provisions of ASCE/SEI 7.9 are used to determine if a load due to snow sliding off the upper roof on to the lower roof must be considered.

Sliding snow loads must be considered because (1) the upper roof is slippery with a slope greater than ¼ on 12 (1.19 degrees) and (2) for separated structures, $h/s = 12.0/9.5 = 1.3 > 1$ and $s = 9.5$ ft < 15 ft (in S.I.: $h/s = 3.7/2.9 = 1.3 > 1$ and $s = 2.9$ m < 4.6 m).

The sliding snow load is equal to the following (see Fig. 3.19):

$$p_{\text{sliding}} = 0.4p_f W/15 = (0.4 \times 59.3 \times 53.0)/15 = 83.8 \text{ lb/ft}^2$$

$$p_{\text{sliding}} = 0.4p_f W/4.6 = (0.4 \times 2.84 \times 16.1)/4.6 = 3.98 \text{ kN/m}^2$$

where p_f for the upper roof is used to determine the sliding snow load (see Example 3.8 for the determination of p_f for the upper roof).

The sliding snow load acts over a width equal to $15 - 9.5 = 5.5$ ft (1.7 m).

The total snow load is equal to the balanced snow load plus the sliding snow load:

$$p_{\text{total}} = p_s + (0.4p_f W/15) = 59.3 + 83.8 = 143.1 \text{ lb/ft}^2$$

$$p_{\text{total}} = p_s + (0.4p_f W/4.6) = 2.84 + 3.98 = 6.82 \text{ kN/m}^2$$

Step 11—Consider rain-on-snow surcharge loads　　　　　　　Sec. 3.14

A rain-on-snow surcharge load of 5 lb/ft² (0.24 kN/m²) is required for locations where (1) the ground snow load, p_g, is 20 lb/ft² (0.96 kN/m²) or less and greater than zero and (2) the roof slope is less than $W/50$ (ASCE/SEI 7.10).

In this example, $p_g = 70$ lb/ft² (3.35 kN/m²), which is greater than 20 lb/ft² (0.96 kN/m²), so a rain-on-snow load is not required.

Step 12—Consider ponding instability   Sec. 3.15

   Because the secondary members are parallel to the free-draining edge of the
roof and the roof slope is less than 1 inch per foot (4.76 degrees), the bay is
susceptible to ponding, and progressive roof deflection and ponding instability
from snow meltwater must be investigated (see Fig. 2.8).

Step 13—Consider snow loads on existing roofs   Sec. 3.16

   Not applicable.

Step 14—Consider snow loads on open-frame equipment structures   Sec. 3.17

   Not applicable.

   The balanced, leeward drift, and sliding snow loads on the lower roof of
the building in this example are depicted in Fig. 3.38.

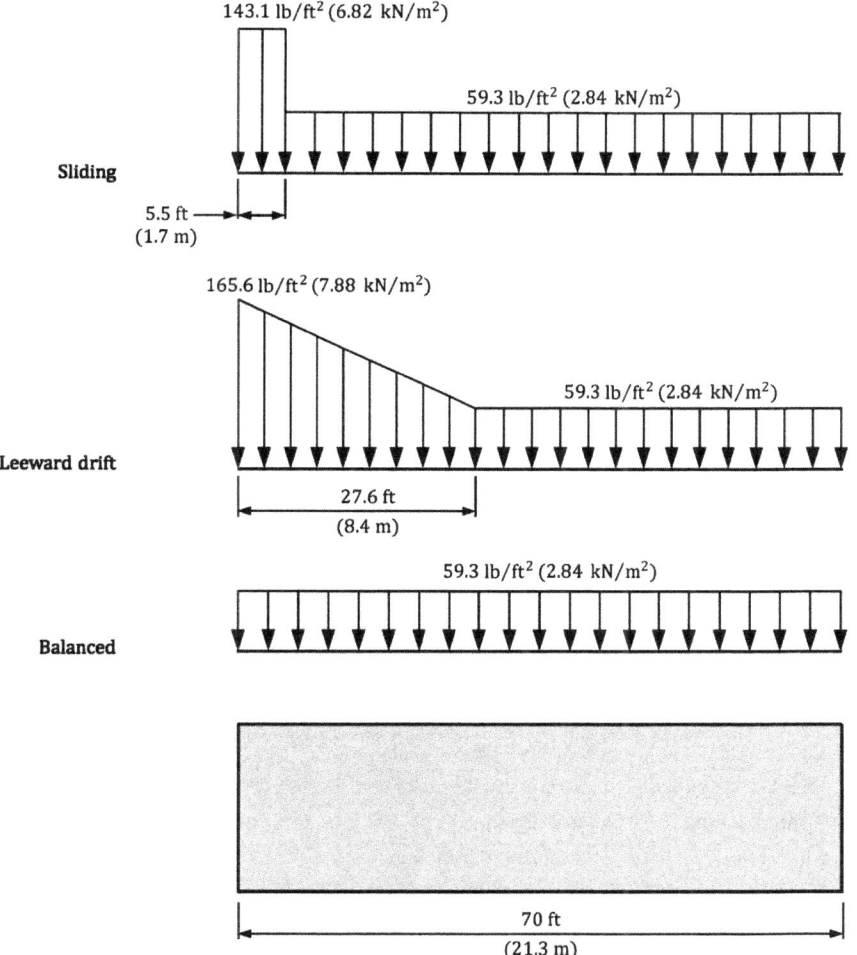

Figure 3.38 Balanced, leeward drift, and sliding snow loads for the lower roof in Example 3.9.

3.18.10   Example 3.10—Calculation of Design Snow Loads for a Building with a Canopy

Determine the design snow loads on the roof and the canopy of the office building in Fig. 3.39 given the design data in Table 3.14.

Solution

Snow Loads on Roof

Step 1—*Determine the ground snow load, p_g*   ASCE/SEI Table 7.2-5

The ground snow load is equal to 39 lb/ft² (1.87 kN/m²).

Step 2—*Determine the flat roof snow load, p_f*   Fig. 3.2

- Step 2a—Determine the surface roughness category

From the design data, the surface roughness category is given as C.

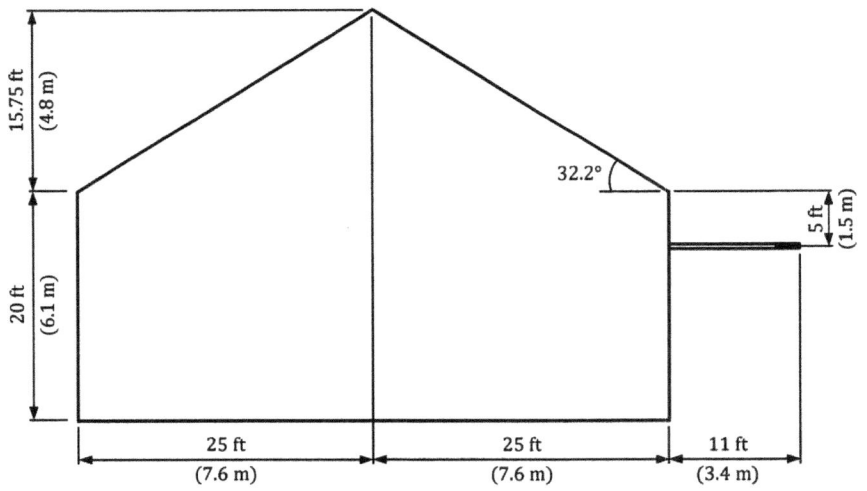

FIGURE 3.39   Elevation of the office building in Example 3.10.

Location	Spokane, WA
Surface roughness	C
Occupancy	Office
Thermal condition	Cold, ventilated roof with $R > 25$ ft²h°F/Btu (4.4 m²K/W)
Roof exposure	Fully exposed
Roof surface	Smooth bituminous
Roof obstructions	None
Roof framing	Simply supported primary and secondary members with the secondary members spanning parallel to the free-draining edges of the roof

TABLE 3.14   Design Data for Example 3.10

- Step 2b—Determine the exposure of the roof

  From the design data, the roof exposure is given as fully exposed.

- Step 2c—Determine the exposure factor, C_e

  Given a surface roughness category of C and a fully exposed roof exposure, $C_e = 0.9$ from ASCE/SEI Table 7.3-1.

- Step 2d—Determine the thermal factor, C_t

  From the design data, the building has a cold, ventilated roof with $R >$ 25 ft²h°F/Btu (4.4 m²K/W). Therefore, $C_t = 1.1$ from ASCE/SEI Table 7.3-2.

- Step 2e—Determine the Risk Category of the building

  From the design data, the building has an office occupancy. Therefore, the Risk Category is II from ASCE/SEI Table 1.5-1.

- Step 2f—Determine the importance factor, I_s

  For a Risk Category II building, $I_s = 1.0$ from ASCE/SEI Table 1.5-2.

- Step 2g—Determine the flat roof snow load, p_f

$$p_f = 0.7C_eC_tI_sp_g = 0.7 \times 0.9 \times 1.1 \times 1.0 \times 39.0 = 27.0 \text{ lb/ft}^2 \qquad (3.1)$$

$$p_f = 0.7C_eC_tI_sp_g = 0.7 \times 0.9 \times 1.1 \times 1.0 \times 1.87 = 1.30 \text{ kN/m}^2$$

Step 3—Determine the minimum snow load for low-slope roofs, p_m   Sec. 3.6

A minimum snow load applies to gable roofs with slopes less than 15 degrees (see Fig. 3.3). Because the upper roof slope in this example is equal to 32.2 degrees, minimum roof snow loads need not be considered.

Step 4—Determine the sloped roof (balanced) snow load, p_s   Sec. 3.7

- Step 4a—Determine the slope factor, C_s Fig. 3.5

  It is determined in Step 2d that $C_t = 1.1$ (cold roof).

  From the design data, there are no obstructions inhibiting the snow from sliding off the roof. Also, because the roof is cold, ice dams need not be considered (if an ice dam can form on a warm roof, it is considered to be an obstruction).

  From the design data, the roof surface is smooth bituminous. According to ASCE/SEI 7.4, smooth bituminous surfaces are considered to be slippery surfaces.

  Because the roof is unobstructed and slippery, the dashed line in ASCE/SEI Figure 7.4-1b can be used to determine C_s. In lieu of using the graph in ASCE/SEI Figure 7.4-1b, use the following equation given in Fig. 3.5, which is the equation of the dashed line in ASCE/SEI Figure 7.4-1b for roof slopes between 10 degrees and 70 degrees, inclusive:

$$C_s = 1.0 - \frac{\text{slope} - 10°}{60°} = 1.0 - \frac{32.2° - 10°}{60°} = 0.63$$

- Step 4b—Determine the sloped roof (balanced) snow load, p_s Fig. 3.9

$$p_s = C_sp_f = 0.63 \times 27.0 = 17.0 \text{ lb/ft}^2 \qquad (3.2)$$

$$p_s = C_sp_f = 0.63 \times 1.30 = 0.82 \text{ kN/m}^2$$

Step 5—Consider loads due to ice dams   Sec. 3.8

Because the roof has been determined to be a cold roof, ice dams and the accompanying uniform load need not be considered (ASCE/SEI 7.4.5).

Step 6—Consider partial loading   Sec. 3.9

From the design data, the primary and secondary framing members are simply supported, so partial loading need not be considered for these members (ASCE/SEI 7.5).

Step 7—Consider unbalanced snow loads   Sec. 3.10

Unbalanced snow loads need not be considered for gable roofs with a slope exceeding 30.3 degrees or with a slope less than 2.39 degrees (ASCE/SEI 7.6.1). The roof slope in this example is 32.2 degrees, which means unbalanced loads need not be considered.

Step 8—Consider drifts on lower roofs   Sec. 3.11

See the section of this solution below for snow loads on the canopy.

Step 9—Consider drifts on roof projections and parapets   Sec. 3.12

Not applicable.

Step 10—Consider sliding snow loads   Sec. 3.13

See the section of this solution below for snow loads on the canopy.

Step 11—Consider rain-on-snow surcharge loads   Sec. 3.14

A rain-on-snow surcharge load of 5 lb/ft^2 (0.24 kN/m^2) is required for locations where (1) the ground snow load, p_g, is 20 lb/ft^2 (0.96 kN/m^2) or less and greater than zero and (2) the roof slope is less than $W/50$ (ASCE/SEI 7.10).

In this example, p_g = 39 lb/ft^2 (1.87 kN/m^2), which is greater than 20 lb/ft^2 (0.96 kN/m^2), so a rain-on-snow load is not required.

Step 12—Consider ponding instability   Sec. 3.15

Because the secondary members are parallel to the free draining edges of the roof and the roof slope is greater than 1 inch per foot (4.76 degrees), the bay is not susceptible to ponding, and progressive roof deflection and ponding instability from snow meltwater need not be investigated (see Fig. 2.8).

Step 13—Consider snow loads on existing roofs   Sec. 3.16

Not applicable.

Step 14—Consider snow loads on open-frame equipment structures   Sec. 3.17

Not applicable.

Snow Loads on Canopy

Step 1—Determine the ground snow load, p_g   ASCE/SEI Table 7.2-5

The ground snow load is equal to 39 lb/ft^2 (1.87 kN/m^2).

Step 2—Determine the flat roof snow load, p_f   Fig. 3.2

- Step 2a—Determine the surface roughness category

From the design data, the surface roughness category is given as C.

- Step 2b—Determine the exposure of the canopy

    Assume the canopy is partially exposed due to the presence of the building.

- Step 2c—Determine the exposure factor, C_e

    Given a surface roughness category of C and a partially exposed roof exposure, $C_e = 1.0$ from ASCE/SEI Table 7.3-1.

- Step 2d—Determine the thermal factor, C_t

    The canopy is an open-air structure. Therefore, $C_t = 1.2$ from ASCE/SEI Table 7.3-2.

- Step 2e—Determine the Risk Category

    From the design data, the building has an office occupancy. Therefore, the Risk Category is II from ASCE/SEI Table 1.5-1.

- Step 2f—Determine the importance factor, I_s

    For a Risk Category II building, $I_s = 1.0$ from ASCE/SEI Table 1.5-2.

- Step 2g—Determine the flat roof snow load, p_f

$$p_f = 0.7 C_e C_t I_s p_g = 0.7 \times 1.0 \times 1.2 \times 1.0 \times 39.0 = 32.8 \text{ lb/ft}^2 \tag{3.1}$$

$$p_f = 0.7 C_e C_t I_s p_g = 0.7 \times 1.0 \times 1.2 \times 1.0 \times 1.87 = 1.57 \text{ kN/m}^2$$

Step 3—Determine the minimum snow load for low-slope roofs, p_m   Sec. 3.6

Because the canopy is essentially flat, a minimum roof snow load must be considered:

For $p_g = 39.0 \text{ lb/ft}^2 > 20.0 \text{ lb/ft}^2$, $p_m = 20 I_s = 20.0 \times 1.0 = 20.0 \text{ lb/ft}^2$ Table 3.3

For $p_g = 1.87 \text{ kN/m}^2 > 0.96 \text{ kN/m}^2$, $p_m = 0.96 I_s = 0.96 \times 1.0 = 0.96 \text{ kN/m}^2$

Step 4—Determine the balanced snow load, p_s   Sec. 3.7

- Step 4a—Determine the slope factor, C_s Fig. 3.5

    It is determined in Step 2d that $C_t = 1.2$ (open-air structure).

    Assuming there are no obstructions inhibiting the snow from sliding off the canopy and the surface of the canopy is smooth bituminous, which is considered to be a slippery surface, use the dashed line in ASCE/SEI Figure 7.4-1c to determine C_s:

    For a flat canopy, $C_s = 1.0$.

- Step 4b—Determine the balanced snow load, p_s

$$p_s = C_s p_f = 1.0 \times 32.8 = 32.8 \text{ lb/ft}^2 \tag{3.2}$$

$$p_s = C_s p_f = 1.0 \times 1.57 = 1.57 \text{ kN/m}^2$$

Step 5—Consider loads due to ice dams   Sec. 3.8

    Not applicable.

Step 6—Consider partial loading   Sec. 3.9

  Not applicable.

Step 7—Consider unbalanced snow loads   Sec. 3.10

  Not applicable.

Step 8—Consider drifts on lower roofs and projections   Sec. 3.11

  Use Fig. 3.15 to determine the leeward and windward drifts forming on the canopy.

- Step 8a—Determine C_s

  From Step 4a, $C_s = 1.0$.

- Step 8b—Determine p_s

  From Step 4b, $p_s = 32.8$ lb/ft^2.

- Step 8c—Determine γ

$$\gamma = 0.13p_g + 14 = (0.13 \times 39.0) + 14 = 19.1 \text{ lb/ft}^3 < 30.0 \text{ lb/ft}^3 \qquad (3.5)$$

- Step 8d—Determine h_b

$$h_b = p_s/\gamma = 32.8/19.1 = 1.7 \text{ ft}$$

- Step 8e—Determine h_c

$$h_c = h_{step} - h_b = 5.0 - 1.7 = 3.3 \text{ ft}$$

- Step 8f—Determine if drift loads must be considered

  Drift loads must be considered where $h_c/h_b \geq 0.2$.

$$h_c/h_b = 3.3/1.7 = 1.9 > 0.2$$

  Therefore, drift loads must be considered.

- Step 8g—Determine the leeward drift height, $h_{d,leeward}$

  Use the following equation, which is applicable where $\ell_{upper} \geq 20$ ft: Table 3.5

$$h_{d,leeward} = \sqrt{I_s} \left\{ [0.43(\ell_{upper})^{1/3}(p_g + 10)^{1/4}] - 1.5 \right\}$$

$$= \sqrt{1.0} \times \left\{ [0.43 \times (50.0)^{1/3} \times (39.0 + 10)^{1/4}] - 1.5 \right\}$$

$$= 2.7 \text{ ft} < 0.6\ell_{lower} = 0.6 \times 11.0 = 6.6 \text{ ft}$$

- Step 8h—Determine the windward drift height, $h_{d,windward}$

  Use the following equation, which is applicable where $\ell_{lower} < 20$ ft:

$$h_{d,windward} = 0.75\sqrt{I_s} \left\{ [0.43(20)^{1/3}(p_g + 10)^{1/4}] - 1.5 \right\} \qquad \text{Table 3.5}$$

$$= 0.75 \times \sqrt{1.0} \times \left\{ [0.43 \times (20)^{1/3} \times (39.0 + 10)^{1/4}] - 1.5 \right\}$$

$$= 1.2 \text{ ft} < \sqrt{I_s p_g \ell_{lower}/4\gamma} = \sqrt{(1.0 \times 39.0 \times 11.0)/(4 \times 19.1)} = 2.4 \text{ ft}$$

- Step 8h—Determine h_d

$$h_d = \text{larger of } (h_{d,\text{leeward}}, h_{d,\text{windward}}) = 2.7 \text{ ft} < h_c = 3.3 \text{ ft}$$

- Step 8i—Determine the drift width, w
  Because $h_d = 2.7 \text{ ft} < h_c = 3.3 \text{ ft}$, $w = 4h_d = 10.8 \text{ ft}$
- Step 8j—Determine the drift load, p_d

$$p_d = \gamma h_d = 19.1 \times 2.7 = 51.6 \text{ lb/ft}^2 \tag{3.8}$$

- Step 8k—Determine the total snow load at the step

$$p_{\text{total}} = p_s + p_d = 32.8 + 51.6 = 84.4 \text{ lb/ft}^2$$

In S.I.:

- Step 8a—Determine C_s
  From Step 4a, $C_s = 1.0$.
- Step 8b—Determine p_s
  From Step 4b, $p_s = 1.57 \text{ kN/m}^2$.
- Step 8c—Determine γ

$$\gamma = 0.426p_g + 2.2 = (0.426 \times 1.87) + 2.2 = 3.0 \text{ kN/m}^3 < 4.7 \text{ kN/m}^3 \qquad \text{Eq. (3.6)}$$

- Step 8d—Determine h_b

$$h_b = p_s / \gamma = 1.57/3.0 = 0.5 \text{ m}$$

- Step 8e—Determine h_c

$$h_c = h_{\text{step}} - h_b = 1.5 - 0.5 = 1.0 \text{ m}$$

- Step 8f—Determine if drift loads must be considered
  Drift loads must be considered where $h_c/h_b \geq 0.2$.

$$h_c/h_b = 1.0/0.5 = 2.0 > 0.2$$

Therefore, drift loads must be considered.

- Step 8g—Determine the leeward drift height, $h_{d,\text{leeward}}$
  Use the following equation, which is applicable where $\ell_{\text{upper}} \geq 6.1$ m:

$$h_{d,\text{leeward}} = \sqrt{I_s} \left\{ [0.42(\ell_{\text{upper}})^{1/3}(p_g + 0.479)^{1/4}] - 0.457 \right\} \qquad \text{Table 3.5}$$

$$= \sqrt{1.0} \times \left\{ [0.42 \times (15.2)^{1/3} \times (1.87 + 0.479)^{1/4}] - 0.457 \right\}$$

$$= 0.83 \text{ m} < 0.6\ell_{\text{lower}} = 0.6 \times 3.4 = 2.0 \text{ m}$$

- Step 8h—Determine the windward drift height, $h_{d,\text{windward}}$

  Use the following equation, which is applicable where $\ell_{\text{lower}} < 6.1$ m:

  $$h_{d,\text{windward}} = 0.75\sqrt{I_s}\left\{[0.42(6.1)^{1/3}(p_g + 0.479)^{1/4}] - 0.457\right\} \qquad \text{Table 3.5}$$

  $$= 0.75 \times \sqrt{1.0} \times \left\{[0.42 \times (6.1)^{1/3} \times (1.87 + 0.479)^{1/4}] - 0.457\right\}$$

  $$= 0.37 \text{ m} < \sqrt{I_s p_g \ell_{\text{lower}}/4\gamma} = \sqrt{(1.0 \times 1.87 \times 3.4)/(4 \times 3.0)} = 0.73 \text{ m}$$

- Step 8h—Determine h_d

  $$h_d = \text{larger of } (h_{d,\text{leeward}}, h_{d,\text{windward}}) = 0.83 \text{ m} < h_c = 1.0 \text{ m}$$

- Step 8i—Determine the drift width, w

  Because $h_d = 0.83 \text{ m} < h_c = 1.0 \text{ m}$, $w = 4h_d = 3.3 \text{ m}$

- Step 8j—Determine the drift load, p_d

  $$p_d = \gamma h_d = 3.0 \times 0.83 = 2.49 \text{ kN/m}^2 \qquad (3.8)$$

- Step 8k—Determine the total snow load at the step

  $$p_{\text{total}} = p_s + p_d = 1.57 + 2.49 = 4.06 \text{ kN/m}^2$$

Step 9—Consider drifts on roof projections and parapets   Sec. 3.12

Not applicable.

Step 10—Consider sliding snow loads   Sec. 3.13

The provisions of ASCE/SEI 7.9 are used to determine if a load due to snow sliding off the roof on to the canopy must be considered.

Loads caused by snow sliding must be considered because the upper roof is slippery with a slope greater than ¼ on 12 (1.19 degrees).

Because $\ell_{\text{lower}} = 11$ ft (3.4 m) < 15 ft (4.6 m), the sliding snow load is permitted to be reduced proportionally (see ASCE/SEI 7.9):

$$p_{\text{sliding}} = 0.4p_f W(\ell_{\text{lower}}/15)/\ell_{\text{lower}} = (0.4 \times 27.0 \times 25.0) \times (11/15)/11 = 18.0 \text{ lb/ft}^2$$

$$p_{\text{sliding}} = 0.4p_f W(\ell_{\text{lower}}/4.6)/\ell_{\text{lower}} = (0.4 \times 1.30 \times 7.6) \times (3.4/4.6)/3.4 = 0.86 \text{ kN/m}^2$$

where p_f of the roof is used to calculate p_{sliding}.

The total snow load over the 11-ft (3.4-m) width is equal to the balanced snow load on the canopy plus sliding snow load from the roof:

$$p_{\text{total}} = p_s + [0.4p_f W(\ell_{\text{lower}}/15)/\ell_{\text{lower}}] = 32.8 + 18.0 = 50.8 \text{ lb/ft}^2$$

$$p_{\text{total}} = p_s + [0.4p_f W(\ell_{\text{lower}}/4.6)/\ell_{\text{lower}}] = 1.57 + 0.86 = 2.43 \text{ kN/m}^2$$

The depth of snow for the total snow load $= p_{\text{total}}/\gamma = 50.8/19.1 = 2.7$ ft (in S.I.: $p_{\text{total}}/\gamma = 2.43/3.0 = 0.81$ m). This depth is less than the distance from the eave of the roof to the top of the canopy, which is equal to $h_{\text{step}} = 5$ ft (1.5 m). Therefore, sliding snow is not blocked and the full load can be developed over the 11-ft (3.4-m) width.

Step 11—Consider rain-on-snow surcharge loads Sec. 3.14

A rain-on-snow surcharge load of 5 lb/ft² (0.24 kN/m²) is required for locations where (1) the ground snow load, p_g, is 20 lb/ft² (0.96 kN/m²) or less and greater than zero and (2) the roof slope is less than $W/50$ (ASCE/SEI 7.10).

In this example, p_g = 39 lb/ft² (1.87 kN/m²), which is greater than 20 lb/ft² (0.96 kN/m²), so a rain-on-snow load is not required.

Step 12—Consider ponding instability Sec. 3.15

Not applicable.

Step 13—Consider snow loads on existing roofs Sec. 3.16

Not applicable.

Step 14—Consider snow loads on open-frame equipment structures Sec. 3.17

Not applicable.

The balanced snow load on the roof and the balanced, drift, and sliding snow loads on the canopy in this example are depicted in Fig. 3.40.

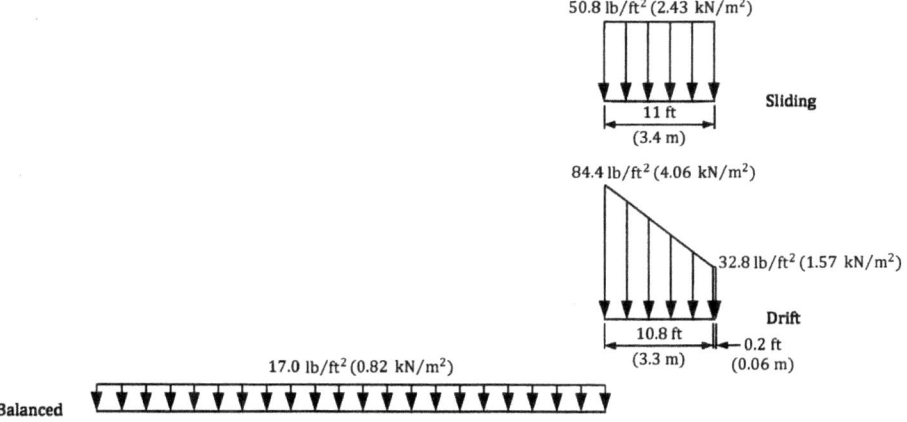

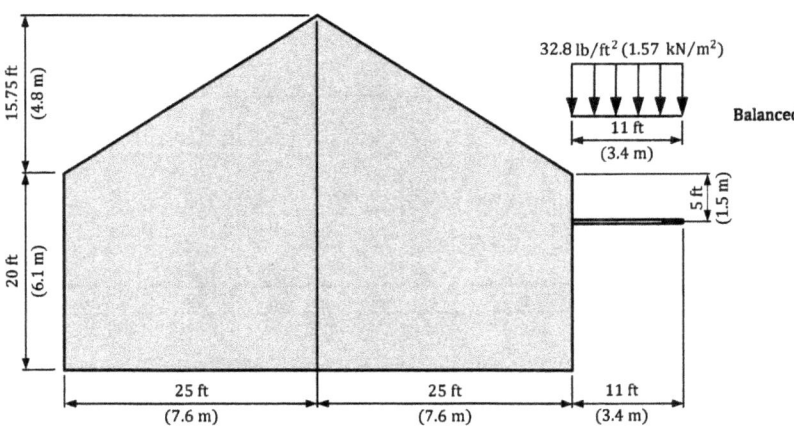

Figure 3.40 Design snow loads for the roof and canopy in Example 3.10.

3.18.11 Example 3.11—Calculation of Design Snow Loads for a Building with a Parapet

Determine the design snow loads on the roof of the fire station in Fig. 3.41 given the design data in Table 3.15.

Solution

Step 1—Determine the ground snow load, p_g   ASCE/SEI Figure. 7.2-1

The ground snow load is equal to 25 lb/ft² (1.20 kN/m²).

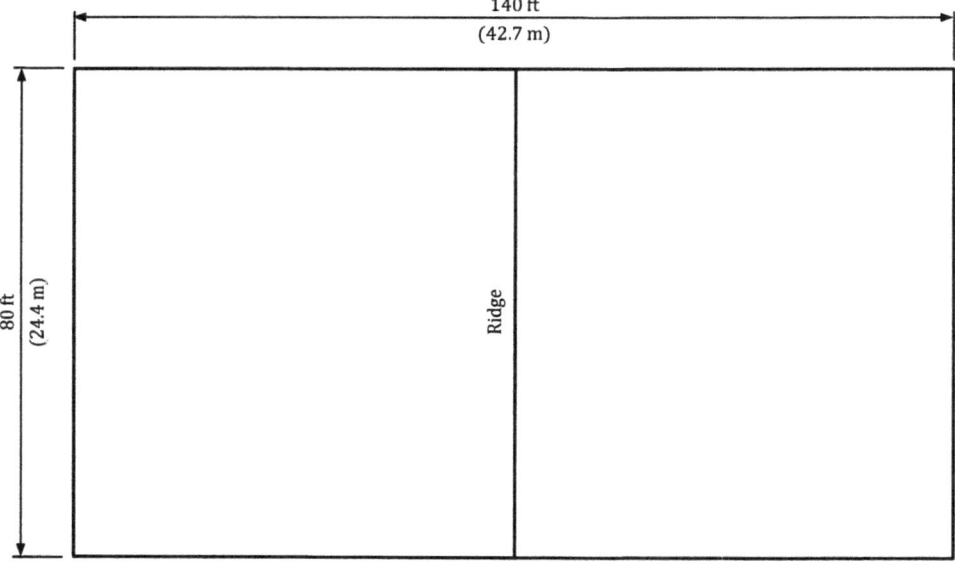

Plan

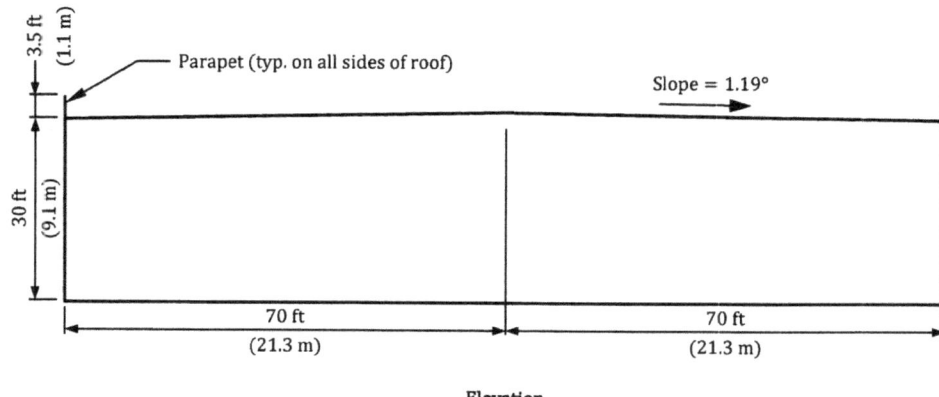

Elevation

FIGURE 3.41 Plan and elevation of the fire station in Example 3.11.

Location	Chicago, IL
Surface roughness	B
Occupancy	Fire station
Thermal condition	Cold, ventilated roof with $R > 25$ ft^{2}h°F/Btu (4.4 m^{2}K/W)
Roof exposure	Partially exposed
Roof surface	Smooth bituminous
Roof obstructions	Parapet on four sides of the building
Roof framing	Simply supported primary and secondary members with the secondary members spanning perpendicular to the edges of the roof

TABLE 3.15 Design Data for Example 3.11

Step 2—Determine the flat roof snow load, p_f Fig. 3.2

- Step 2a—Determine the surface roughness category

  From the design data, the surface roughness category is given as B.

- Step 2b—Determine the exposure of the roof

  From the design data, the roof exposure is given as partially exposed.

- Step 2c—Determine the exposure factor, C_e

  Given a surface roughness category of B and a partially exposed roof exposure, $C_e = 1.0$ from ASCE/SEI Table 7.3-1.

- Step 2d—Determine the thermal factor, C_t

  From the design data, the building has a cold, ventilated roof with $R > 25$ ft^{2}h°F/Btu (4.4 m^{2}K/W). Therefore, $C_t = 1.1$ from ASCE/SEI Table 7.3-2.

- Step 2e—Determine the Risk Category of the building

  From the design data, the building is a fire house, which designated as an essential facility. Therefore, the Risk Category is IV from ASCE/SEI Table 1.5-1.

- Step 2f—Determine the importance factor, I_s

  For a Risk Category IV building, $I_s = 1.2$ from ASCE/SEI Table 1.5-2.

- Step 2g—Determine the flat roof snow load, p_f

$$p_f = 0.7 C_e C_t I_s p_g = 0.7 \times 1.0 \times 1.1 \times 1.2 \times 25.0 = 23.1 \text{ lb/ft}^2 \quad (3.1)$$

$$p_f = 0.7 C_e C_t I_s p_g = 0.7 \times 1.0 \times 1.1 \times 1.2 \times 1.20 = 1.11 \text{ kN/m}^2$$

Step 3—Determine the minimum snow load for low-slope roofs, p_m Sec. 3.6

A minimum snow load applies to gable roofs with slopes less than 15 degrees (see Fig. 3.3). Because the upper roof slope in this example is equal to 1.19 degrees, minimum roof snow loads must be considered:

For $p_g = 25.0$ lb/ft$^2 > 20.0$ lb/ft^2, $p_m = 20 I_s = 20.0 \times 1.2 = 24.0$ lb/ft^2 Table 3.3

For $p_g = 1.20$ kN/m$^2 > 0.96$ kN/m^2, $p_m = 0.96 I_s = 0.96 \times 1.2 = 1.15$ kN/m^2

Step 4—Determine the sloped roof (balanced) snow load, p_s   Sec. 3.7

- Step 4a—Determine the slope factor, C_s Fig. 3.5

  It is determined in Step 2d that $C_t = 1.1$ (cold roof).

  From the design data, the parapets inhibit the snow from sliding off the roof. This constitutes obstructions on all sides of the roof.

  From the design data, the roof surface is smooth bituminous. According to ASCE/SEI 7.4, smooth bituminous surfaces are considered to be slippery surfaces.

  Because the roof is obstructed and slippery, the solid line in ASCE/SEI Figure 7.4-1b is used to determine C_s:

  For a roof slope of 1.19 degrees, which is less than 37.5 degrees, $C_s = 1.0$.

- Step 4b—Determine the sloped roof (balanced) snow load, p_s Fig. 3.9

$$p_s = C_s p_f = 1.0 \times 23.1 = 23.1 \text{ lb/ft}^2 \tag{3.2}$$

$$p_s = C_s p_f = 1.0 \times 1.11 = 1.11 \text{ kN/m}^2$$

Step 5—Consider loads due to ice dams   Sec. 3.8

Because the roof has been determined to be a cold roof, ice dams and the accompanying uniform load need not be considered (ASCE/SEI 7.4.5).

Step 6—Consider partial loading   Sec. 3.9

From the design data, the primary and secondary framing members are simply supported, so partial loading need not be considered for these members (ASCE/SEI 7.5).

Step 7—Consider unbalanced snow loads   Sec. 3.10

Unbalanced snow loads need not be considered for gable roofs with a slope exceeding 30.3 degrees or with a slope less than 2.39 degrees (ASCE/SEI 7.6.1). The roof slope in this example is 1.19 degrees, which means unbalanced loads need not be considered.

Step 8—Consider drifts on lower roofs   Sec. 3.11

Not applicable.

Step 9—Consider drifts on roof projections and parapets   Sec. 3.12

Drift loads on parapets are determined using the provisions of ASCE/SEI 7.7.1 (ASCE/SEI 7.8).

Windward drifts occur at parapet walls and Fig. 3.18 is used to determine the windward drift load:

- Step 9a—Determine C_s

  From Step 4a, $C_s = 1.0$.

- Step 9b—Determine p_s

  From Step 4b, $p_s = 23.1$ lb/ft^2.

- Step 9c—Determine γ

$$\gamma = 0.13 p_g + 14 = (0.13 \times 25.0) + 14 = 17.3 \text{ lb/ft}^3 < 30.0 \text{ lb/ft}^3 \tag{3.5}$$

- Step 9d—Determine h_b

$$h_b = p_s / \gamma = 23.1/17.3 = 1.3 \text{ ft}$$

- Step 9e—Determine h_c

$$h_c = h_{step} - h_b = 3.5 - 1.3 = 2.2 \text{ ft}$$

- Step 9f—Determine if drift loads must be considered
  Drift loads must be considered where $h_c / h_b \geq 0.2$.

$$h_c / h_b = 2.2/1.3 = 1.7 > 0.2$$

Therefore, drift loads must be considered.

- Step 9g—Determine the windward drift height, h_d
  Use the following equation based on the largest upwind fetch $\ell_u = 140$ ft:

$$h_d = 0.75\sqrt{I_s} \left\{ [0.43(\ell_u)^{1/3}(p_g + 10)^{1/4}] - 1.5 \right\}$$

$$= 0.75 \times \sqrt{1.2} \times \left\{ [0.43 \times (140.0)^{1/3} \times (25.0 + 10)^{1/4}] - 1.5 \right\}$$

$$= 3.2 \text{ ft} > h_c = 2.2 \text{ ft}$$

- Step 9h—Determine the drift width, w

Because $h_d = 3.2$ ft $> h_c = 2.2$ ft, $w = 4h_d^2 / h_c = 18.6$ ft $> 8h_c = 17.6$ ft

Use $w = 17.6$ ft.

- Step 9i—Determine the drift load, p_d

$$p_d = \gamma h_c = 17.3 \times 2.2 = 38.1 \text{ lb/ft}^2$$

- Step 9j—Determine the total snow load at the face of the parapet

$$p_{total} = \gamma h_{step} = 17.3 \times 3.5 = 60.6 \text{ lb/ft}^2$$

In S.I.:

- Step 9a—Determine C_s
  From Step 4a, $C_s = 1.0$.
- Step 9b—Determine p_s
  From Step 4b, $p_s = 1.11 \text{ kN/m}^2$.
- Step 9c—Determine γ

$$\gamma = 0.426p_g + 2.2 = (0.426 \times 1.20) + 2.2 = 2.7 \text{ kN/m}^3 < 4.7 \text{ kN/m}^3 \qquad (3.6)$$

- Step 9d—Determine h_b

$$h_b = p_s/\gamma = 1.11/2.7 = 0.41 \text{ m}$$

- Step 9e—Determine h_c

$$h_c = h_{step} - h_b = 1.1 - 0.41 = 0.69 \text{ m}$$

- Step 9f—Determine if drift loads must be considered

Drift loads must be considered where $h_c/h_b \geq 0.2$.

$$h_c/h_b = 0.69/0.41 = 1.7 > 0.2$$

Therefore, drift loads must be considered.

- Step 9g—Determine the windward drift height, h_d

Use the following equation based on the largest upwind fetch $\ell_u = 42.7 \text{ m}$:

$$h_d = 0.75\sqrt{I_s}\left\{[0.42(\ell_u)^{1/3}(p_g + 0.479)^{1/4}] - 0.457\right\}$$

$$= 0.75 \times \sqrt{1.2} \times \left\{[0.42 \times (42.7)^{1/3} \times (1.20 + 0.479)^{1/4}] - 0.457\right\}$$

$$= 1.0 \text{ m} > h_c = 0.69 \text{ m}$$

- Step 9h—Determine the drift width, w

Because $h_d = 1.0 \text{ m} > h_c = 0.69 \text{ m}$, $w = 4h_d^2/h_c = 5.8 \text{ m} > 8h_c = 5.5 \text{ m}$

Use $w = 5.5 \text{ m}$.

- Step 9i—Determine the drift load, p_d

$$p_d = \gamma h_c = 2.7 \times 0.69 = 1.86 \text{ kN/m}^2$$

- Step 9j—Determine the total snow load at the face of the parapet

$$p_{total} = \gamma h_{step} = 2.7 \times 1.1 = 2.97 \text{ kN/m}^2$$

Step 10—Consider sliding snow loads   Sec. 3.13

Not applicable.

Step 11—Consider rain-on-snow surcharge loads   Sec. 3.14

A rain-on-snow surcharge load of 5 lb/ft² (0.24 kN/m²) is required for locations where (1) the ground snow load, p_g, is 20 lb/ft² (0.96 kN/m²) or less and greater than zero and (2) the roof slope is less than $W/50$ (ASCE/SEI 7.10).

In this example, $p_g = 25$ lb/ft² (1.20 kN/m²), which is greater than 20 lb/ft² (0.96 kN/m²), so a rain-on-snow load is not required.

Step 12—Consider ponding instability   Sec. 3.15

Because the secondary members are perpendicular to the edges of the roof and the roof slope is equal to 1.19 degrees, the bay is not susceptible to ponding, and progressive roof deflection and ponding instability from snow meltwater need not be investigated (see Fig. 2.7).

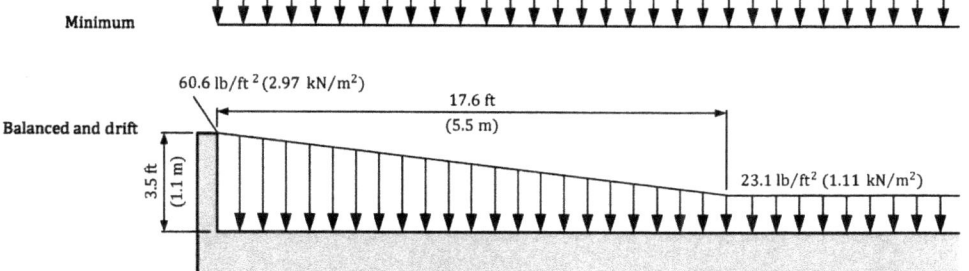

FIGURE 3.42 Design snow loads for the roof in Example 3.11.

Step 13—Consider snow loads on existing roofs   Sec. 3.16

Not applicable.

Step 14—Consider snow loads on open-frame equipment structures   Sec. 3.17

Not applicable.

The minimum, balanced, and drift snow loads on the roof are depicted in Fig. 3.42.

3.18.12 Example 3.12—Calculation of Design Snow Loads for a Building with a Rooftop Unit

Determine the design snow loads on the roof of the fire station in Fig. 3.43. Use the design data in Table 3.15 of Example 3.11. In this example, the roof has no parapets.

Solution

Steps 1 through 8—Results are the same as in Example 3.11

$$p_g = 25 \text{ lb/ft}^2 \ (1.20 \text{ kN/m}^2)$$

$$C_e = 1.0$$

$$C_t = 1.1$$

$$I_s = 1.2$$

$$p_f = p_s = 23.1 \text{ lb/ft}^2 \ (1.11 \text{ kN/m}^2)$$

$$p_m = 24.0 \text{ lb/ft}^2 \ (1.15 \text{ kN/m}^2)$$

$$\gamma = 17.3 \text{ lb/ft}^3 \ (2.7 \text{ kN/m}^3)$$

$$h_b = 1.3 \text{ ft} \ (0.41 \text{ m})$$

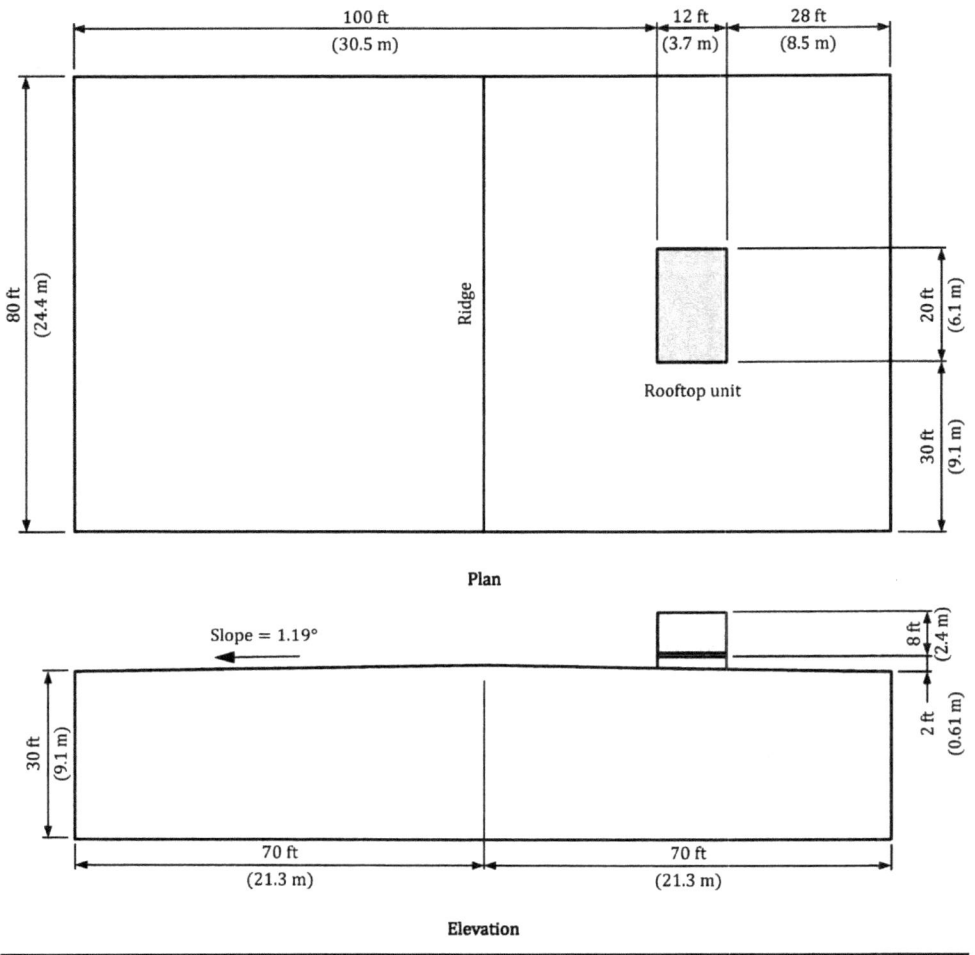

100 ft (30.5 m)   12 ft (3.7 m)   28 ft (8.5 m)

80 ft (24.4 m)

Ridge

Rooftop unit

20 ft (6.1 m)

30 ft (9.1 m)

Plan

Slope = 1.19°

8 ft (2.4 m)

2 ft (0.61 m)

30 ft (9.1 m)

70 ft (21.3 m)   70 ft (21.3 m)

Elevation

FIGURE 3.43 Plan and elevation of the fire station in Example 3.12.

Step 9—Consider drifts on roof projections and parapets   Sec. 3.12

Drift loads on projections are determined using the provisions of ASCE/SEI 7.7.1 (ASCE/SEI 7.8).

Windward drifts occur at roof projections and Fig. 3.18 is used to determine the windward drift load.

- Step 9a—Determine h_c

$$h_c = h_{step} - h_b = 10.0 - 1.3 = 8.7 \text{ ft}$$

- Step 9b—Determine if drift loads must be considered

Drift loads must be considered where $h_c/h_b \geq 0.2$.

$$h_c/h_b = 8.7/1.3 = 6.7 > 0.2$$

Therefore, drift loads must be considered.

Drift loads need not be applied to the sides of the rooftop unit less than 15 ft (ASCE/SEI 7.8). Therefore, drift loads are not required on the 12-ft-long sides of the rooftop unit.

Also, drift loads need not be applied where $H \geq 2$ ft.

In this case, $H = 2.0 - h_b = 2.0 - 1.3 = 0.7$ ft < 2 ft, so drift loads must be applied on the 20-ft-long sides of the rooftop unit.

- Step 9c—Determine the windward drift height, h_d

The larger of the upwind fetches perpendicular to the 20-ft-long sides of the rooftop unit is equal to the larger clear distance to the face of the unit plus one-half the width of the rooftop unit in that direction:

$$\ell_u = 100.0 + (12.0/2) = 106.0 \text{ ft}$$

For simplicity, this fetch is used to determine h_d on both sides of the rooftop unit:

$$h_d = 0.75\sqrt{I_s}\left\{[0.43(\ell_u)^{1/3}(p_g + 10)^{1/4}] - 1.5\right\}$$

$$= 0.75 \times \sqrt{1.2} \times \left\{[0.43 \times (106.0)^{1/3} \times (25.0 + 10)^{1/4}] - 1.5\right\}$$

$$= 2.8 \text{ ft} < h_c = 8.7 \text{ ft}$$

- Step 9d—Determine the drift width, w

Because $h_d = 2.8$ ft $< h_c = 8.7$ ft, $w = 4h_d = 11.2$ ft

- Step 9e—Determine the drift load, p_d

$$p_d = \gamma h_d = 17.3 \times 2.8 = 48.4 \text{ lb/ft}^2$$

- Step 9f—Determine the total snow load at the face of the rooftop unit

$$p_{\text{total}} = p_s + p_d = 23.1 + 48.4 = 71.5 \text{ lb/ft}^2$$

In S.I.:

- Step 9a—Determine h_c

$$h_c = h_{\text{step}} - h_b = 3.01 - 0.41 = 2.6 \text{ m}$$

- Step 9b—Determine if drift loads must be considered

Drift loads must be considered where $h_c/h_b \geq 0.2$.

$$h_c/h_b = 2.6/0.41 = 6.3 > 0.2$$

Therefore, drift loads must be considered.

Drift loads need not be applied to the sides of the rooftop unit less than 4.6 m (ASCE/SEI 7.8). Therefore, drift loads are not required on the 3.7-m-long sides of the rooftop unit.

Also, drift loads need not be applied where $H \geq 0.61$ m.

In this case, $H = 0.61 - h_b = 0.61 - 0.4 = 0.21$ m < 0.61 m, so drift loads must be applied on the 6.1-m-long sides of the rooftop unit.

- Step 9c—Determine the windward drift height, h_d

  The larger of the upwind fetches perpendicular to the 6.1-m-long sides of the rooftop unit is equal to the larger clear distance to the face of the unit plus one-half the width of the rooftop unit in that direction:

  $$\ell_u = 30.5 + (3.7/2) = 32.4 \text{ m}$$

  For simplicity, this fetch is used to determine h_d on both sides of the rooftop unit:

  $$h_d = 0.75\sqrt{I_s}\left\{[0.42(\ell_u)^{1/3}(p_g + 0.479)^{1/4}] - 0.457\right\}$$

  $$= 0.75 \times \sqrt{1.2} \times \left\{[0.42 \times (32.4)^{1/3} \times (1.20 + 0.479)^{1/4}] - 0.457\right\}$$

  $$= 0.88 \text{ m} < h_c = 2.6 \text{ m}$$

- Step 9d—Determine the drift width, w

  Because $h_d = 0.88$ m $< h_c = 2.6$ m, $w = 4h_d = 3.5$ m

- Step 9e—Determine the drift load, p_d

  $$p_d = \gamma h_d = 2.7 \times 0.88 = 2.38 \text{ kN/m}^2$$

- Step 9f—Determine the total snow load at the face of the rooftop unit

  $$p_{total} = p_s + p_d = 1.11 + 2.38 = 3.49 \text{ kN/m}^2$$

Step 10—Consider sliding snow loads   Sec. 3.13

Not applicable.

Step 11—Consider rain-on-snow surcharge loads   Sec. 3.14

A rain-on-snow surcharge load of 5 lb/ft² (0.24 kN/m²) is required for locations where (1) the ground snow load, p_g, is 20 lb/ft² (0.96 kN/m²) or less and greater than zero and (2) the roof slope is less than $W/50$ (ASCE/SEI 7.10).
In this example, $p_g = 25$ lb/ft² (1.20 kN/m²), which is greater than 20 lb/ft² (0.96 kN/m²), so a rain-on-snow load is not required.

Step 12—Consider ponding instability   Sec. 3.15

Because the secondary members are perpendicular to the free-draining edges of the roof and the roof slope is equal to 1.19 degrees, the bay is not susceptible to ponding, and progressive roof deflection and ponding instability from snow meltwater need not be investigated (see Fig. 2.8).

Step 13—Consider snow loads on existing roofs   Sec. 3.16

Not applicable.

Step 14—Consider snow loads on open-frame equipment structures   Sec. 3.17

Not applicable.
The minimum, balanced, and drift snow loads on the roof are depicted in Fig. 3.44. Although not specifically addressed in ASCE/SEI 7.8, the balanced snow load is applied to the roof beneath the area of the rooftop unit.

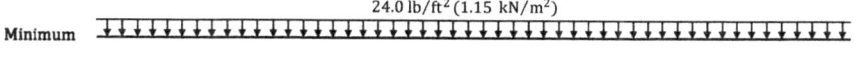

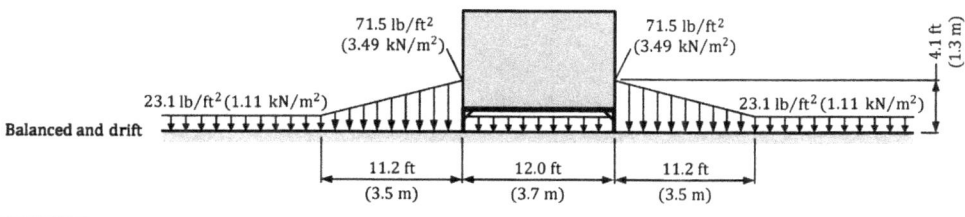

FIGURE 3.44 Design snow loads for the roof in Example 3.12.

3.18.13 Example 3.13—Calculation of Design Snow Loads for an Existing Building Adjacent to a New Building

Determine the design snow loads on the roof of the existing commercial building in Fig. 3.45 before and after the construction of the new adjacent commercial warehouse building given the design data in Table 3.16.

Solution

Snow Loads Prior to Construction of New Building

Step 1—Determine the ground snow load, p_g   ASCE/SEI Table 7.2-7

The ground snow load is equal to 11 lb/ft² (0.53 kN/m²).

Step 2—Determine the flat roof snow load, p_f   Fig. 3.2

- Step 2a—Determine the surface roughness category

  From the design data, the surface roughness category is given as B.

- Step 2b—Determine the exposure of the roof

  From the design data, the roof exposure is given as partially exposed.

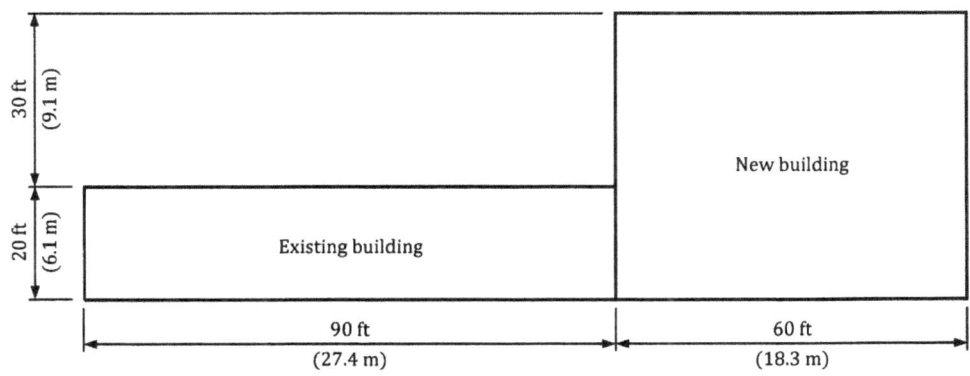

FIGURE 3.45 Elevations of the existing and new commercial buildings in Example 3.13.

Location	Portland, OR
Surface roughness	B
Occupancy	Commercial
Thermal condition — Existing	Unheated
Thermal condition — New	Structure kept just above freezing
Roof exposures	Partially exposed
Roof surfaces	Metal
Roof obstructions	None
Roof framing — Existing	Reinforced-concrete wide-module joist system with joists spaced 5 ft (1.5 m) on center and 30 ft (9.1 m) spans
Roof framing — New	Simply supported primary and secondary members with the secondary members spanning perpendicular to the free draining edges of the roof
Roof slope	Both roofs are nominally flat except for localized areas around roof drains sloped to facilitate drainage

TABLE 3.16 Design Data for Example 3.13

- Step 2c—Determine the exposure factor, C_e

  Given a surface roughness category of B and a partially exposed roof exposure, $C_e = 1.0$ from ASCE/SEI Table 7.3-1.

- Step 2d—Determine the thermal factor, C_t

  From the design data, the existing building is unheated. Therefore, $C_t = 1.2$ from ASCE/SEI Table 7.3-2.

- Step 2e—Determine the Risk Category of the building

  From the design data, the existing building has a commercial occupancy. Therefore, the Risk Category is II from ASCE/SEI Table 1.5-1.

- Step 2f—Determine the importance factor, I_s

  For a Risk Category II building, $I_s = 1.0$ from ASCE/SEI Table 1.5-2.

- Step 2g—Determine the flat roof snow load, p_f

$$p_f = 0.7C_eC_tI_sp_g = 0.7\times1.0\times1.2\times1.0\times11.0 = 9.2 \text{ lb/ft}^2 \qquad (3.1)$$

$$p_f = 0.7C_eC_tI_sp_g = 0.7\times1.0\times1.2\times1.0\times0.53 = 0.45 \text{ kN/m}^2$$

Step 3—Determine the minimum snow load for low-slope roofs, p_m　　　　Sec. 3.6

Because the existing roof is essentially flat, minimum roof snow loads must be considered:

For $p_g = 11.0 \text{ lb/ft}^2 < 20.0 \text{ lb/ft}^2$, $p_m = I_sp_g = 1.0\times11.0 = 11.0 \text{ lb/ft}^2$　　　Table 3.3

For $p_g = 0.53 \text{ kN/m}^2 < 0.96 \text{ kN/m}^2$, $p_m = I_sp_g = 1.0\times0.53 = 0.53 \text{ kN/m}^2$

Step 4—Determine the sloped roof (balanced) snow load, p_s Sec. 3.7

• Step 4a—Determine the slope factor, C_s Fig. 3.5

It is determined in Step 2d that $C_t = 1.2$ (cold roof).

From the design data, there are no obstructions inhibiting the snow from sliding off the existing roof and the roof surface is metal. According to ASCE/SEI 7.4, metal surfaces are considered to be slippery surfaces.

Because the roof is unobstructed and slippery, the dashed line in ASCE/SEI Figure 7.4-1c is used to determine C_s:

For a flat roof, $C_s = 1.0$.

• Step 4b—Determine the sloped roof (balanced) snow load, p_s Fig. 3.9

$$p_s = C_s p_f = 1.0 \times 9.2 = 9.2 \text{ lb/ft}^2 \quad\quad (3.2)$$
$$p_s = C_s p_f = 1.0 \times 0.45 = 0.45 \text{ kN/m}^2$$

Step 5—Consider loads due to ice dams Sec. 3.8

Because the roof has been determined to be a cold roof, ice dams and the accompanying uniform load need not be considered (ASCE/SEI 7.4.5).

Step 6—Consider partial loading Sec. 3.9

From the design data, the reinforced-concrete joists are continuous beams. Therefore, partial loading must be considered. The three load cases defined in ASCE/SEI 7.5.1 are given in Fig. 3.46 for a typical reinforced-concrete joist, which has a tributary width of 5 ft (1.5 m).

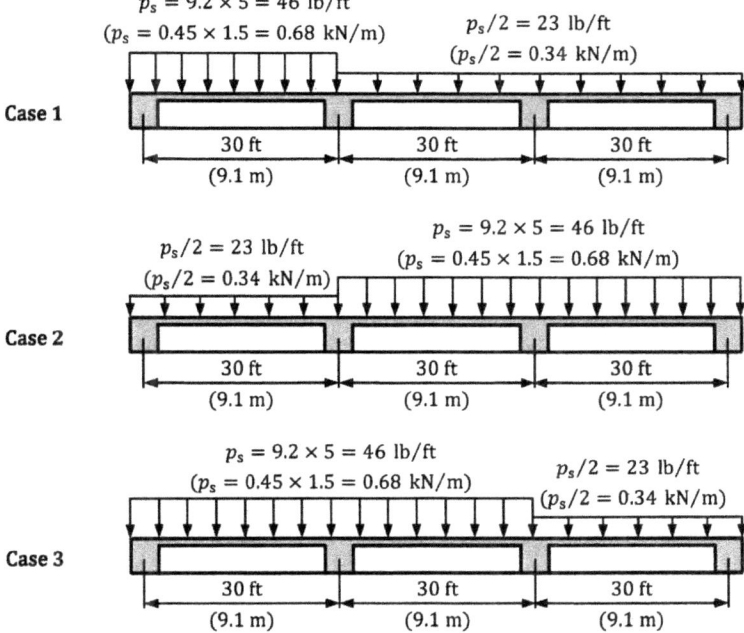

FIGURE 3.46 Partial loading cases for a reinforced-concrete joist in Example 3.13.

Step 7—Consider unbalanced snow loads   Sec. 3.10

   Unbalanced snow loads need not be considered for flat roofs.

Step 8—Consider drifts on lower roofs   Sec. 3.11

   Not applicable.

Step 9—Consider drifts on roof projections and parapets   Sec. 3.12

   Not applicable.

Step 10—Consider sliding snow loads   Sec. 3.13

   Not applicable.

Step 11—Consider rain-on-snow surcharge loads   Sec. 3.14

   A rain-on-snow surcharge load of 5 lb/ft² (0.24 kN/m²) is required for locations where (1) the ground snow load, p_g, is 20 lb/ft² (0.96 kN/m²) or less and greater than zero and (2) the roof slope is less than $W/50$ (ASCE/SEI 7.10).

   In this example, $p_g = 11$ lb/ft² (0.53 kN/m²), which is less than 20 lb/ft² (0.96 kN/m²), so a rain-on-snow load is required. An additional 5 lb/ft² (0.24 kN/m²) must be added to the balanced snow load of 9.2 lb/ft² (0.45 kN/m²). This additional load need not be used in combination with the partial loading determined in Step 6.

   For a typical reinforced-concrete joist, the balanced snow load is equal to the following:

$$p_s = (9.2 + 5.0) \times 5.0 = 71.0 \text{ lb/ft}$$

$$p_s = (0.45 + 0.24) \times 1.5 = 1.04 \text{ kN/m}$$

The balanced snow load on a typical reinforced-concrete joist is depicted in Fig. 3.47.

Step 12—Consider ponding instability   Sec. 3.15

   Because the roof is essentially flat, the bays are susceptible to ponding, and progressive roof deflection and ponding instability from snow meltwater must be investigated.

Step 13—Consider snow loads on existing roofs   Sec. 3.16

   Not applicable.

Step 14—Consider snow loads on open-frame equipment structures   Sec. 3.17

   Not applicable.

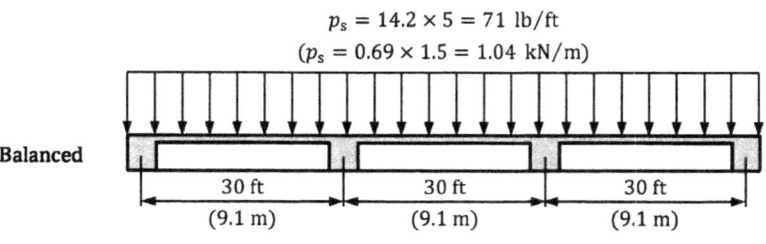

$$p_s = 14.2 \times 5 = 71 \text{ lb/ft}$$
$$(p_s = 0.69 \times 1.5 = 1.04 \text{ kN/m})$$

Balanced

30 ft   30 ft   30 ft
(9.1 m)   (9.1 m)   (9.1 m)

FIGURE 3.47   Balanced snow load for a reinforced-concrete joist in Example 3.13.

Snow Loads after Construction of New Building   The snow loads on the existing roof prior to construction of the new building are also applicable after construction of the new building. Because of the presence of the new building, drift loads on the existing roof must also be considered. Sliding snow loads need not be considered because the upper roof is essentially flat.

Step 8—Consider drifts on lower roofs   Sec. 3.11

Use Fig. 3.15 to determine the leeward and windward drifts forming on the lower roof.

- Step 8a—Determine γ

$$\gamma = 0.13 p_g + 14 = (0.13 \times 11.0) + 14 = 15.4 \text{ lb/ft}^3 < 30.0 \text{ lb/ft}^3 \qquad (3.5)$$

- Step 8b—Determine h_b

$$h_b = p_s/\gamma = 9.2/15.4 = 0.60 \text{ ft}$$

- Step 8c—Determine h_c

$$h_c = h_{step} - h_b = 30.0 - 0.60 = 29.4 \text{ ft}$$

- Step 8d—Determine if drift loads must be considered
Drift loads must be considered where $h_c/h_b \geq 0.2$.

$$h_c/h_b = 29.4/0.60 = 49.0 > 0.2$$

Therefore, drift loads must be considered.

- Step 8e—Determine the leeward drift height, $h_{d,\text{leeward}}$
Use the following equation, which is applicable where $\ell_{\text{upper}} \geq 20$ ft:

$$h_{d,\text{leeward}} = \sqrt{I_s}\left\{[0.43(\ell_{\text{upper}})^{1/3}(p_g+10)^{1/4}]-1.5\right\} \leq 0.6\ell_{\text{lower}} \qquad \text{Table 3.5}$$

$$= \sqrt{1.0} \times \left\{[0.43 \times (60.0)^{1/3} \times (11.0+10)^{1/4}]-1.5\right\}$$

$$= 2.1 \text{ ft} < 0.6 \times 90.0 = 54.0 \text{ ft}$$

- Step 8f—Determine the windward drift height, $h_{d,\text{windward}}$
Use the following equation, which is applicable where $\ell_{\text{lower}} \geq 20$ ft:

$$h_{d,\text{windward}} = 0.75\sqrt{I_s}\left\{[0.43(\ell_{\text{lower}})^{1/3}(p_g+10)^{1/4}]-1.5\right\} \qquad \text{Table 3.5}$$

$$= 0.75 \times \sqrt{1.0} \times \left\{[0.43 \times (90.0)^{1/3} \times (11.0+10)^{1/4}]-1.5\right\} = 2.0 \text{ ft}$$

- Step 8g—Determine h_d

$$h_d = \text{larger of } (h_{d,\text{leeward}}, h_{d,\text{windward}}) = 2.1 \text{ ft}$$

- Step 8h—Determine the drift width, w

  Because $h_d = 2.1$ ft $< h_c = 29.4$ ft, $w = 4h_d = 8.4$ ft

- Step 8i—Determine the drift load, p_d

  $$p_d = \gamma h_d = 15.4 \times 2.1 = 32.3 \ \text{lb/ft}^2$$

- Step 8j—Determine the total snow load at the face of the new building

  $$p_{total} = p_s + p_d = 9.2 + 32.3 = 41.5 \ \text{lb/ft}^2$$

In S.I.:

- Step 8a—Determine γ

  $$\gamma = 0.426 p_g + 2.2 = (0.426 \times 0.53) + 2.2 = 2.4 \ \text{kN/m}^3 < 4.7 \ \text{kN/m}^3 \qquad (3.6)$$

- Step 8b—Determine h_b

  $$h_b = p_s / \gamma = 0.45/2.4 = 0.19 \ \text{m}$$

- Step 8c—Determine h_c

  $$h_c = h_{step} - h_b = 9.1 - 0.19 = 8.9 \ \text{m}$$

- Step 8d—Determine if drift loads must be considered

  Drift loads must be considered where $h_c / h_b \geq 0.2$.

  $$h_c / h_b = 8.9/0.19 = 47.0 > 0.2$$

  Therefore, drift loads must be considered.

- Step 8e—Determine the leeward drift height, $h_{d,\text{leeward}}$

  Use the following equation, which is applicable where $\ell_{\text{upper}} \geq 6.1$ m:

  $$h_{d,\text{leeward}} = \sqrt{I_s} \left\{ [0.42(\ell_{\text{upper}})^{1/3}(p_g + 0.479)^{1/4}] - 0.457 \right\} \leq 0.6 \ell_{\text{lower}} \qquad \text{Table 3.5}$$

  $$= \sqrt{1.0} \times \left\{ [0.42 \times (18.3)^{1/3} \times (0.53 + 0.479)^{1/4}] - 0.457 \right\}$$

  $$= 0.65 \ \text{m} < 0.6 \times 27.4 = 16.4 \ \text{m}$$

- Step 8f—Determine the windward drift height, $h_{d,\text{windward}}$

  Use the following equation, which is applicable where $\ell_{\text{lower}} \geq 6.1$ m:

  $$h_{d,\text{windward}} = 0.75 \sqrt{I_s} \left\{ [0.42(\ell_{\text{lower}})^{1/3}(p_g + 0.479)^{1/4}] - 0.457 \right\} \qquad \text{Table 3.5}$$

  $$= 0.75 \times \sqrt{1.0} \times \left\{ [0.42 \times (27.4)^{1/3} \times (0.53 + 0.479)^{1/4}] - 0.457 \right\} = 0.61 \ \text{m}$$

- Step 8g—Determine h_d

  $$h_d = \text{larger of } (h_{d,\text{leeward}}, h_{d,\text{windward}}) = 0.65 \ \text{m}$$

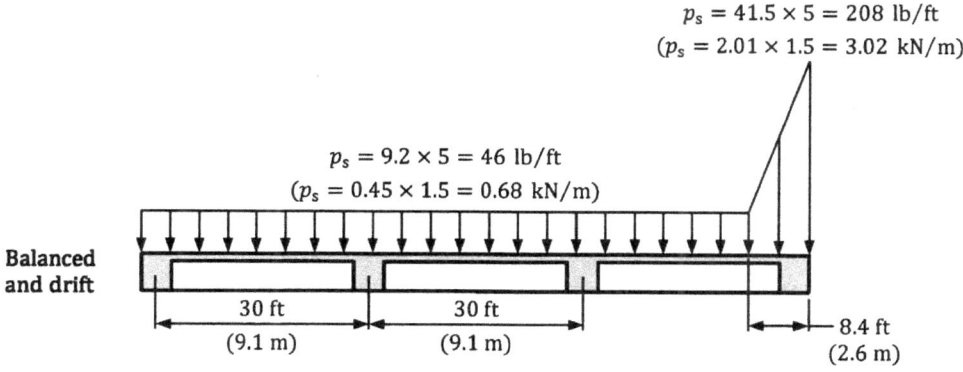

$p_s = 41.5 \times 5 = 208$ lb/ft
$(p_s = 2.01 \times 1.5 = 3.02$ kN/m)

$p_s = 9.2 \times 5 = 46$ lb/ft
$(p_s = 0.45 \times 1.5 = 0.68$ kN/m)

Balanced
and drift

30 ft
(9.1 m)

30 ft
(9.1 m)

8.4 ft
(2.6 m)

Figure 3.48   Balanced and drift snow loads for a reinforced-concrete joist in Example 3.13.

- Step 8h—Determine the drift width, w
  Because $h_d = 0.65$ m $< h_c = 8.9$ m, $w = 4h_d = 2.6$ m
- Step 8i—Determine the drift load, p_d

$$p_d = \gamma h_d = 2.4 \times 0.65 = 1.56 \text{ kN/m}^2$$

- Step 8j—Determine the total snow load at the face of the new building

$$p_{total} = p_s + p_d = 0.45 + 1.56 = 2.01 \text{ kN/m}^2$$

The balanced and drift snow loads on a typical reinforced-concrete joist are depicted in Fig. 3.48.

3.18.14   Example 3.14—Calculation of Design Snow Loads for Elements in an Open-Frame Equipment Structure

Determine the design snow loads for the following elements in an open-frame equipment structure: (a) a 12-in. (305-mm) diameter pipe (which includes the insulation on the pipe) and (b) adjacent cable trays (see Fig. 3.49). Assume the following: (1) $p_g = 30$ lb/ft^2 (1.44 kN/m^2), (2) $C_e = 1.0$, and (3) $I_s = 1.0$.

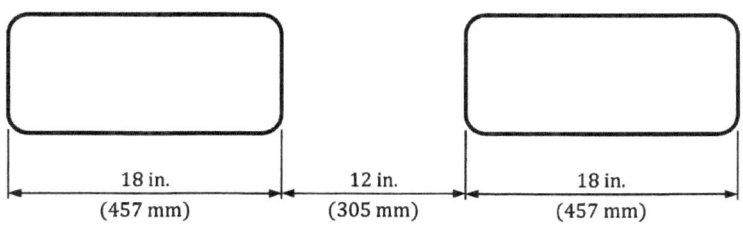

18 in.
(457 mm)

12 in.
(305 mm)

18 in.
(457 mm)

Figure 3.49   Adjacent cable trays in Example 3.14.

Solution

Part (a): 12-in. (305-mm) Diameter Pipe

Step 1—Determine the flat roof snow load, p_f

- Step 1a—Determine the exposure factor, C_e
  From the design data, $C_e = 1.0$.
- Step 1b—Determine the thermal factor, C_t
  For open-frame equipment structures, $C_t = 1.2$.   ASCE/SEI 7.13
- Step 1c—Determine the importance factor, I_s
  From the design data, $I_s = 1.0$.
- Step 1d—Determine the flat roof snow load, p_f

$$p_f = 0.7C_eC_tI_sp_g = 0.7 \times 1.0 \times 1.2 \times 1.0 \times 30.0 = 25.2 \text{ lb/ft}^2 \tag{3.1}$$

$$p_f = 0.7C_eC_tI_sp_g = 0.7 \times 1.0 \times 1.2 \times 1.0 \times 1.44 = 1.21 \text{ kN/m}^2$$

Step 2—Determine γ

$$\gamma = 0.13p_g + 14 = (0.13 \times 30.0) + 14 = 17.9 \text{ lb/ft}^3 < 30.0 \text{ lb/ft}^3 \tag{3.5}$$

$$\gamma = 0.426p_g + 2.2 = (0.426 \times 1.44) + 2.2 = 2.8 \text{ kN/m}^3 < 4.7 \text{ kN/m}^3 \tag{3.6}$$

Step 3—Determine the depth of the flat roof snow load

$$\text{Depth} = p_f/\gamma = 25.2/17.9 = 1.4 \text{ ft} = 16.8 \text{ in.}$$
$$\text{Depth} = p_f/\gamma = 1.21/2.8 = 0.43 \text{ m} = 430 \text{ mm}$$

Step 4—Determine the snow load on the pipe   Table 3.6

$$0.73p_f/\gamma = (0.73 \times 25.2)/17.9 = 1.03 \text{ ft} = 12.4 \text{ in.}$$

Because the diameter of the pipe is less than $0.73p_f/\gamma$, the snow load on the pipe is triangular (see ASCE/SEI Figure 7.13-2a).

Height of the snow on the pipe = $1.37D = 1.37 \times 12.0 = 16.4$ in.

Maximum snow load = $1.37D\gamma = 1.37 \times (12.0/12) \times 17.9 = 24.5$ lb/ft^2

With an assumed angle of repose = 70 degrees, area of snow $A = (12.0 \times 16.4)/2 = 98.4$ in.^{2}

Snow load per length of pipe = $\gamma A = 17.9 \times (98.4/144) = 12.2$ lb/ft

In S.I.:

$$0.73p_f/\gamma = (0.73 \times 1.21)/2.8 = 0.32 \text{ m} = 320 \text{ mm}$$

Because the diameter of the pipe is less than $0.73p_f/\gamma$, the snow load on the pipe is triangular (see ASCE/SEI Figure 7.13-2a).

Height of the snow on the pipe = $1.37D = 1.37 \times 305 = 418$ mm

Maximum snow load = $1.37D\gamma = 1.37 \times 0.305 \times 2.8 = 1.17$ kN/m^2

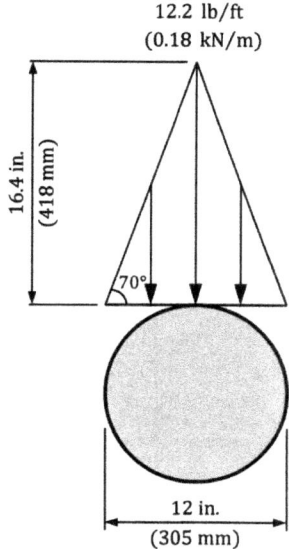

12.2 lb/ft
(0.18 kN/m)

16.4 in.
(418 mm)

70°

12 in.
(305 mm)

Figure 3.50 Design snow load on the pipe in Example 3.14.

With an assumed angle of repose = 70 degrees, area of snow $A = (305 \times 418)/2 = 63{,}745$ mm² $= 0.064$ m²

Snow load per length of pipe $= \gamma A = 2.8 \times 0.064 = 0.18$ kN/m

The design snow load on the pipe is given in Fig. 3.50.

Part (b): Adjacent Cable Trays The 18-in. width of each cable tray is greater than $0.73 p_f / \gamma = 12.4$ in. Therefore, the snow load on each cable tray is trapezoidal with a maximum pressure of $p_f = 25.2$ lb/ft² (see ASCE/SEI Figure 7.13-2b).

Height of the snow on the cable tray $h = p_f / \gamma = 25.2/17.9 = 1.4$ ft $= 16.8$ in.

Clear space between the adjacent cable trays $S_p = 12.0$ in. $< h = 16.8$ in.

Therefore, an additional uniform cornice load of p_f must be applied in the space between the adjacent cable trays.

With an assumed angle of repose = 70 degrees, the area of snow tributary to each cable tray is equal to the following:

$$A = 16.8 \times \left[18.0 + \frac{12.0}{2} - \left(\frac{1}{2} \times \frac{16.8}{\tan 70°} \right) \right] = 351.8 \text{ in.}^2$$

Snow load per cable tray $= \gamma A = 17.9 \times (351.8/144) = 43.7$ lb/ft

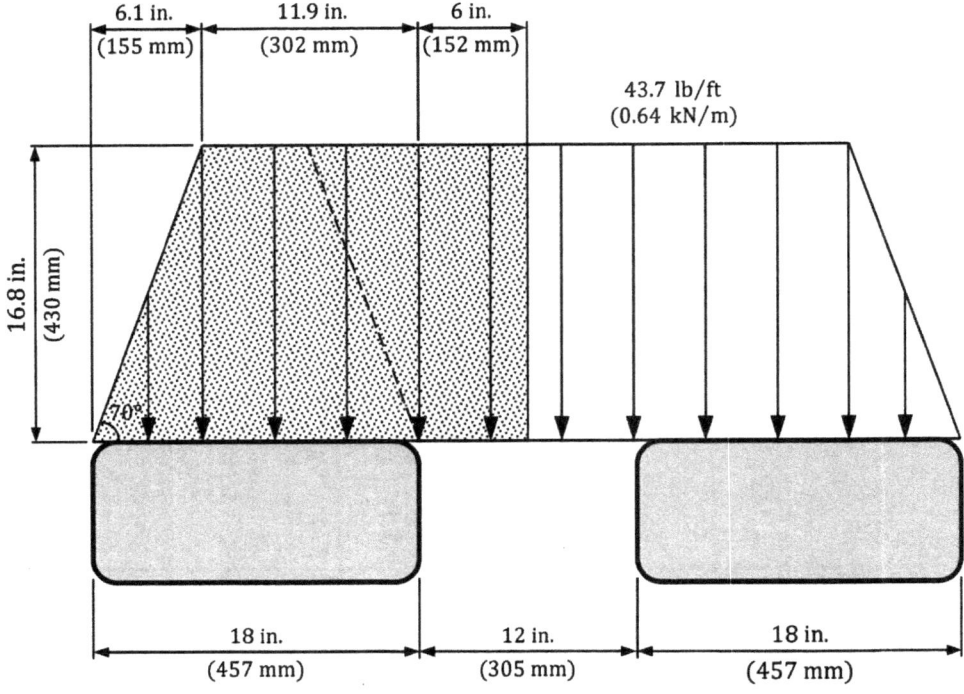

FIGURE 3.51 Design snow load on the cable trays in Example 3.14.

In S.I.:

The 457-mm width of each cable tray is greater than $0.73p_f/\gamma = 320$ mm. Therefore, the snow load on each cable tray is trapezoidal with a maximum pressure of $p_f = 1.21$ kN/m² (see ASCE/SEI Figure 7.13-2b).

Height of the snow on the cable tray $h = p_f/\gamma = 1.21/2.8 = 0.43$ m = 430 mm

Clear space between the adjacent cable trays $S_p = 305$ mm $< h = 430$ mm

Therefore, an additional uniform cornice load of p_f must be applied in the space between the adjacent cable trays.

With an assumed angle of repose = 70 degrees, the area of snow tributary to each cable tray is equal to the following:

$$A = 430 \times \left[457 + \frac{305}{2} - \left(\frac{1}{2} \times \frac{430}{\tan 70°} \right) \right] = 228,436 \text{ mm}^2 = 0.228 \text{ m}^2$$

Snow load per cable tray $= \gamma A = 2.8 \times 0.228 = 0.64$ kN/m

The design snow load on the cable trays is given in Fig. 3.51.

CHAPTER 4

Ice Loads

4.1 Overview

This chapter contains methods to calculate design atmospheric ice loads due to freezing rain in accordance with IBC 1614 and ASCE/SEI Chapter 10. Ice-sensitive structures are structures for which the effect of an atmospheric icing load governs the design of part or all of the structure (IBC 202 and ASCE/SEI 10.2). Examples include the following: (1) lattice structures, (2) guyed masts, (3) overhead lines, (4) light suspension and cable-stayed bridges, (5) aerial cable systems (like those for ski lifts), (6) amusement rides, (7) open catwalks and platforms, (8) flagpoles, and (9) signs.

The provisions of ASCE/SEI Chapter 10 do not apply to structures covered by national standards (see ASCE/SEI 10.1.3 for examples). In such cases, the applicable standards are to be used to determine the ice loads.

4.2 Notation

A_f = projected area normal to the wind except where C_f is specified for the actual surface area, ft² (m²)

A_i = cross-sectional area of ice, in.² (mm²)

A_s = surface area of one side of a flat plate or the projected area of complex shapes, in.² (mm²)

= gross area of a freestanding wall or freestanding solid sign, ft² (m²)

b = width of an angle leg, flat plate, or tube, in. (mm)

b_f = flange width of a W-shape or channel, in. (mm)

C_f = force coefficient to be used in determination of wind loads for other structures

D = diameter of a circular structure or member, ft (m)

D_c = diameter of the cylinder circumscribing an object, in. (mm)

D_i = weight of ice, lb/ft (N/m)

d = overall depth of a W-shape or channel, in. (mm)

f_z = factor to account for the increase in ice thickness with height

G = gust-effect factor

h = depth of an angle leg, flat plate, or tube, in. (mm)

= mean roof height of a building, ft (m)

I_i = importance factor for ice thickness from ASCE/SEI Table 1.5-2 based on the Risk Category from ASCE/SEI Table 1.5-1

I_w = importance factor for concurrent wind pressure from ASCE/SEI Table 1.5-2 based on the Risk Category from ASCE/SEI Table 1.5-1

K_d = wind directionality factor

K_e = ground elevation factor

K_h = velocity pressure exposure coefficient evaluated at height $z = h$

K_z = velocity pressure exposure coefficient evaluated at height z

K_{zt} = topographic factor

n_1 = fundamental natural frequency, Hz

q_h = velocity pressure evaluated at height $z = h$ abound ground, lb/ft^2 (N/m^2)

q_z = velocity pressure evaluated at height z abound ground, lb/ft^2 (N/m^2)

r = radius of the maximum cross-section of a dome or radius of a sphere, ft (m)

t = nominal ice thickness on a cylinder caused by freezing rain at a height of 33 ft (10 m), in. (mm)

t_d = design ice thickness, in. (mm)

t_p = thickness of a flat plate, in. (mm)

V_c = concurrent wind speed, mi/h (m/s)

V_i = volume of ice, in.^{3} (mm^3)

W_i = wind-on-ice load, lb (N)

z = height above ground, ft (m)

z_g = nominal height of the atmospheric boundary layer, ft (m)

α = 3-s gust-speed power law exponent

ε = solidity ratio

4.3   Procedure to Determine Design Atmospheric Ice Loads

A step-by-step procedure to determine design atmospheric ice loads is given in Fig. 4.1 (ASCE/SEI 10.8). The sections of this publication referenced in Fig. 4.1 contain information to determine the design ice thickness, t_d, the ice weight, D_i, and the wind-on-ice load, W_i.

4.4   Ice Loads Caused by Freezing Rain

Ice loads caused by freezing rain are determined using the ice weight, D_i, formed on all exposed surfaces of structural members, guys, components, appurtenances, and cable systems. The following sections provide information needed to calculate D_i.

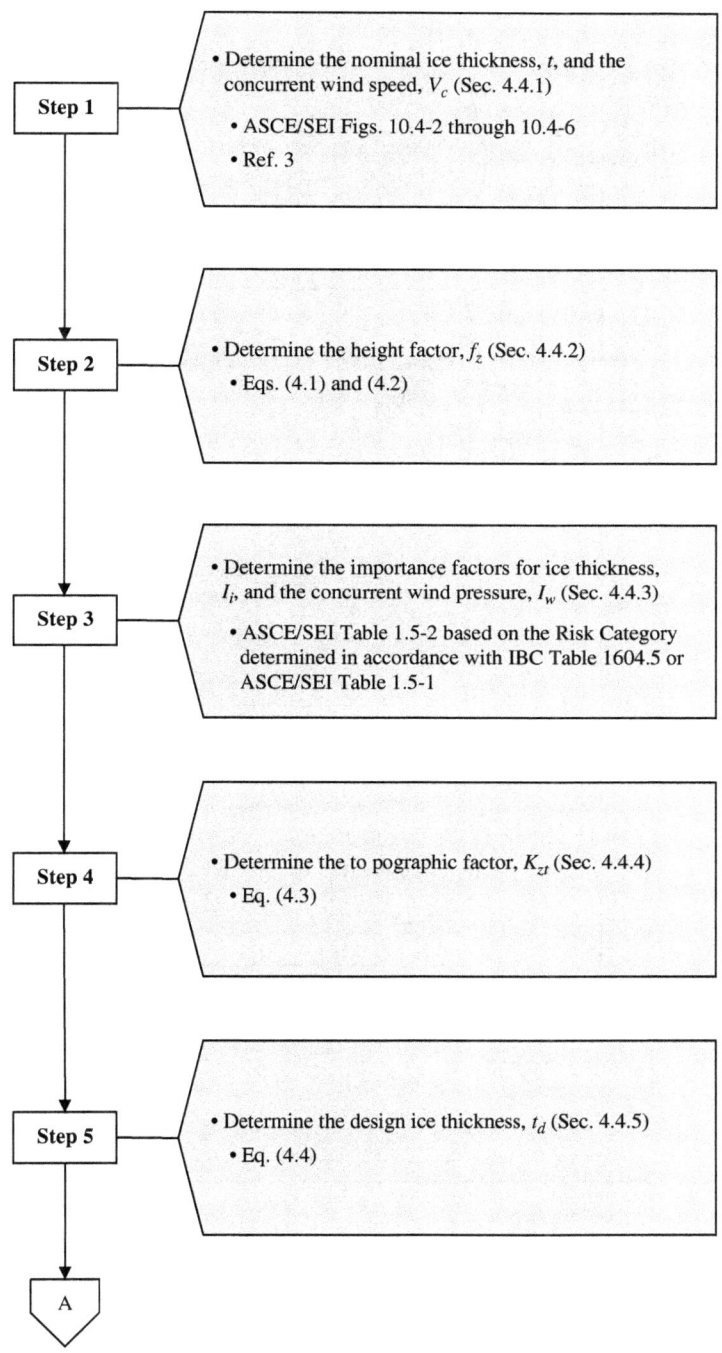

FIGURE 4.1   Procedure to determine design atmospheric ice loads.

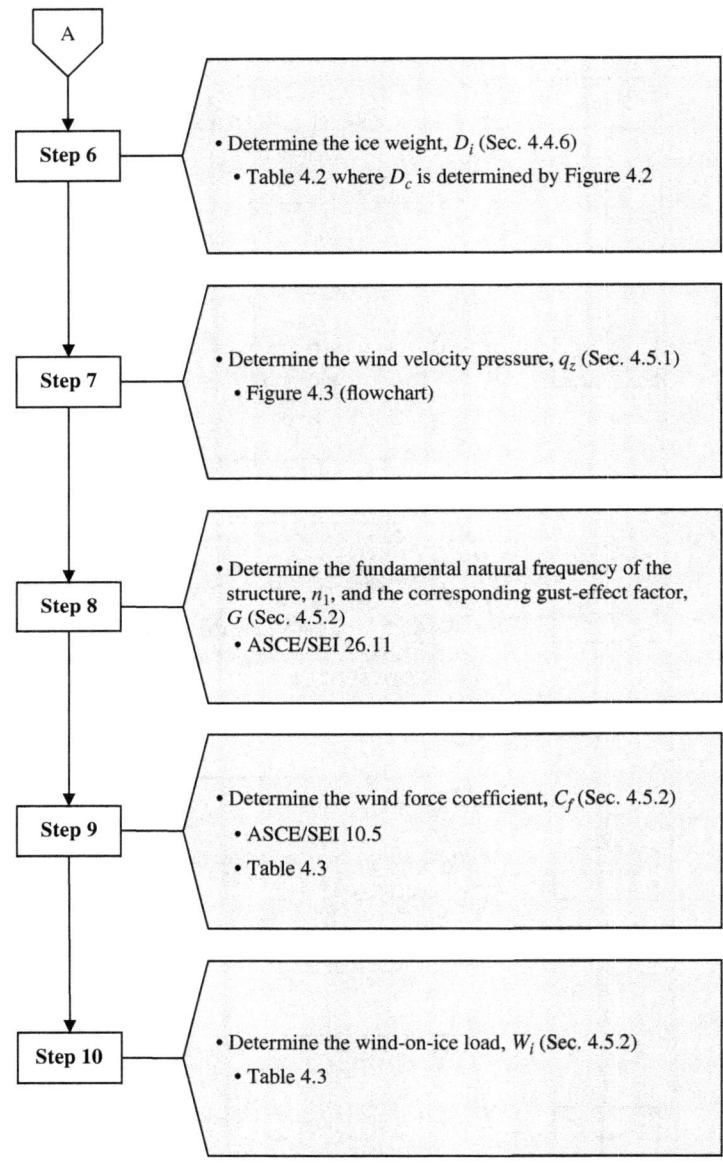

Figure 4.1   (*Continued*)

4.4.1   Nominal Ice Thickness

ASCE/SEI Figures 10.4-2 through 10.4-6 provide an equivalent uniform radial thickness of ice, t, due to freezing rain at a height of 33 ft (10 m) above the ground for the contiguous 48 states and Alaska based on a 500-year mean recurrence interval. Detail maps are provided for Lake Superior, Fraser Valley, and Columbia River Gorge. Also given in the figures are the concurrent 3-second gust wind speeds, V_c, which correspond to the winds occurring during the freezing rainstorm and between the time the freezing rain stops and the time the temperature rises to above freezing. Values of t and V_c for an address or latitude and longitude of a site can be obtained from Ref. 3. Factors to adjust the 500-year ice thickness to one based on other mean recurrence intervals are given in ASCE/SEI C10.4-1.

Special icing regions are also identified on the maps (gray shaded areas) and occur in the western mountainous regions and in the Appalachian Mountains. In the former, ice thicknesses may exceed the mapped value in foothills and passes, while in the latter, ice thicknesses may vary significantly over short distances because of local variations in elevation, topography, and exposure. The ice thicknesses in ASCE/SEI Figure 10.4-2 must be adjusted in mountainous regions to account for both freezing rain and in-cloud icing (see ASCE/SEI C10.4.2). It is good practice to consult with the local building official when making such adjustments.

4.4.2 Height Factor

The height factor, f_z, adjusts the mapped values of t in ASCE/SEI Figures 10.4-2 through 10.4-6, which are based on a height of 33 ft (10 m) above ground, to any height z above ground [ASCE/SEI Equations (10.4-4) and (10.4-4si)]:

$$f_z = \begin{cases} \left(\dfrac{z}{33}\right)^{0.10} & \text{for } 0 \text{ ft} < z \le 900 \text{ ft} \\[2ex] 1.4 \text{ for } z > 900 \text{ ft} \end{cases} \tag{4.1}$$

$$f_z = \begin{cases} \left(\dfrac{z}{10}\right)^{0.10} & \text{for } 0 \text{ m} < z \le 275 \text{ m} \\[2ex] 1.4 \text{ for } z > 275 \text{ m} \end{cases} \tag{4.2}$$

4.4.3 Importance Factor

The ice importance factors for ice thickness, I_i, and concurrent wind pressure, I_w, are given in ASCE/SEI Table 1.5-2 based on the Risk Category of a structure, which is determined by IBC Table 1604.5 or ASCE/SEI Table 1.5-1. The mean recurrence intervals corresponding to the risk categories are given in Table 4.1.

Risk Category	Mean Recurrence Interval (Years)
I	250
II	500
III	1,000
IV	1,400

Table 4.1 Mean Recurrence Intervals Based on Risk Category

4.4.4 Topographic Factor

Because of wind speed-up effects, t and V_c are larger for buildings and structures situated on hills, ridges, and escarpments compared to those located on level terrain. To account for these effects, t is modified by $(K_{zt})^{0.35}$ where K_{zt} is the topographic factor determined by ASCE/SEI Equation (26.8-1):

$$K_{zt} = (1 + K_1 K_2 K_3)^2 \tag{4.3}$$

The topographic multipliers K_1, K_2, and K_3 are determined from ASCE/SEI Figure 26.8-1.

Not every hill, ridge, or escarpment requires an increase in t or V_c; these quantities must be increased only when the site conditions and structure locations of ASCE/SEI 26.8.1 are met. Otherwise, $K_{zt} = 1.0$.

4.4.5 Design Ice Thickness for Freezing Rain

The design ice thickness, t_d, is determined by ASCE/SEI Equation (10.4-5):

$$t_d = t I_i f_z (K_{zt})^{0.35} \quad \text{(in. and mm)} \tag{4.4}$$

This thickness is used in calculating ice weight, D_i (see Sec. 4.4.6).

4.4.6 Ice Weight

For structural shapes, prismatic members and other similar shapes, the weight of ice, D_i, is determined by multiplying the density of ice [which is equal to 56 lb/ft³ (900 kg/m³); see ASCE/SEI 10.4.1] by the cross-sectional area of ice, $A_i = \pi t_d (D_c + t_d)$ [see ASCE/SEI Equation (10.4-1)]:

$$D_i = 56 A_i = 56 \pi t_d (D_c + t_d) \quad \text{(lb/ft)} \tag{4.5}$$

$$D_i = 900 g A_i = 900 \times (9.8/1,000) A_i = 8.82 \pi t_d (D_c + t_d) \quad \text{(kN/m)} \tag{4.6}$$

The term D_c is the diameter of a cylinder circumscribing a shape or member and is given in ASCE/SEI Figure 10.4-1 for different cross-sectional shapes (see Fig. 4.2).

For flat plates and large three-dimensional objects such as domes and spheres, D_i is determined by multiplying the density of ice by the volume of ice, $V_i = \pi t_d A_s$ [see ASCE/SEI Equation (10.4-2)]:

$$D_i = 56 \pi t_d A_s \quad \text{(lb)} \tag{4.7}$$

$$D_i = 8.82 \pi t_d A_s \quad \text{(kN)} \tag{4.8}$$

For vertical plates, A_s is equal to 0.8 times the area of one side of the plate. Similarly, A_s is equal to 0.6 times the area of one side of the plate for horizontal plates. For domes and spheres, $A_s = \pi r^2$ [ASCE/SEI Equation (10.4-3)] where r is the radius of the maximum cross section of a dome or the radius of a sphere.

The equations in Table 4.2 can be used to determine D_i for the shapes and objects noted above.

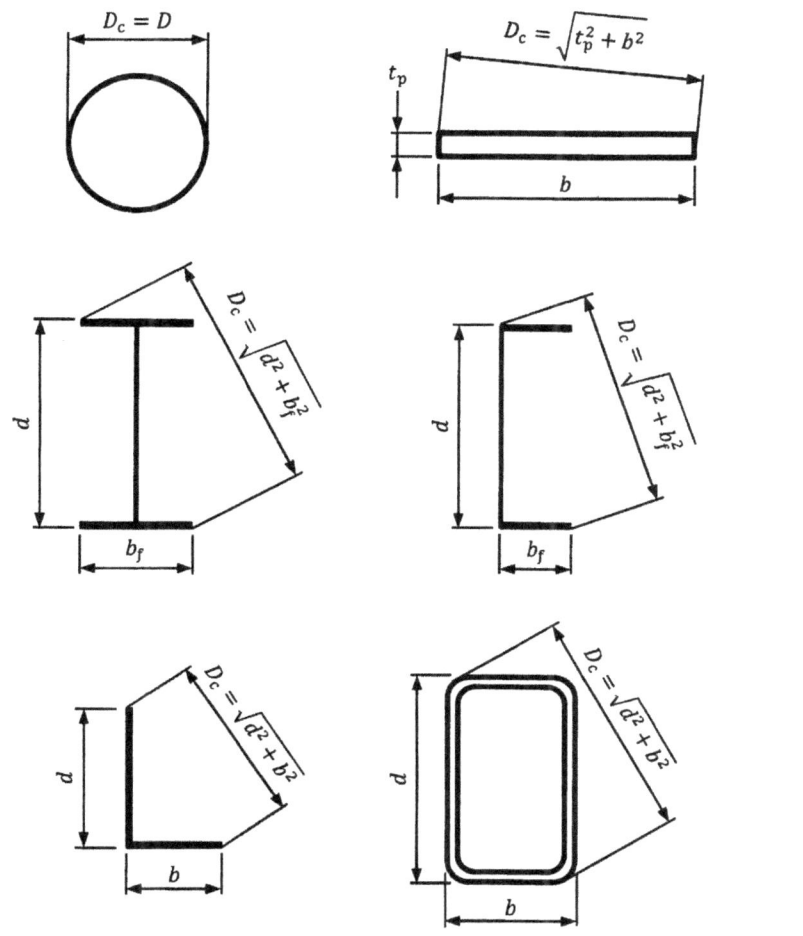

Figure 4.2 Dimension D_c for different cross-sectional shapes.

Shape/Object	D_i
Structural shapes, prismatic members, and other similar shapes	$0.389\pi t_d(D_c + t_d)$ (lb/ft) $(8.82 \times 10^{-3})\pi t_d(D_c + t_d)$ (N/m)
Vertical plates	$3.73\pi t_d bh$ (lb) $7.06\pi t_d bh$ (N)
Horizontal plates	$2.80\pi t_d bh$ (lb) $5.29\pi t_d bh$ (N)
Domes and spheres	$4.67\pi^2 t_d r^2$ (lb) $8.82\pi^2 t_d r^2$ (N)

$^*t_d$ and D_c are in in. (mm).
b, h, and r are in ft (m).

Table 4.2 Ice Weight, D_i, on Shapes and Objects Caused by Freezing Rain*

4.5   Wind on Ice-Covered Structures

4.5.1   Wind Velocity Pressure, q_z

Wind loads on ice-covered structures are determined using the wind velocity pressure, q_z, which is determined by ASCE/SEI Equations (26.10-1) and (26.10-1.si):

$$q_z = 0.00256 K_z K_{zt} K_d K_e V_c^2 \quad (\text{lb/ft}^2) \tag{4.9}$$

$$q_z = 0.613 K_z K_{zt} K_d K_e V_c^2 \quad (\text{N/m}^2) \tag{4.10}$$

where V_c is the concurrent wind speed determined from ASCE/SEI Figures 10.4-2 through 10.4-6 in miles per hour (mi/h) [meters per second (m/s)]. The design wind velocity pressures are equal to q_z determined by Eqs. (4.9) and (4.10) multiplied by the importance factor I_w, which is equal to 1.0 for all risk categories (see ASCE/SE Table 1.5-2).

The flowchart in Fig. 4.3 can be used to determine q_z.

4.5.2   Wind-On-Ice Loads

Ice formed on structural members, components, and appurtenances increases the projected area exposed to wind and changes the structure's wind drag coefficients. The increased projected area is determined by adding t_d to all free edges of the projected area. Ice-sensitive structures must be designed for the wind loads determined by the provisions in ASCE/SEI Chapters 26 through 31 using the increased projected area due to ice formation and the modifications in ASCE/SEI 10.5.1 through 10.5.5. The loads determined using these provisions are defined as wind-on-ice loads, W_i.

Wind-on-ice loads for the structures in ASCE/SEI 10.5.1 through 10.5.5 are given in Table 4.3. The gust-effect factor, G, is determined in accordance with ASCE/SEI 26.11 based on the fundamental natural frequency, n_1, which can be determined by any rational method. The gust-effect factor is permitted to be taken as 0.85 for a rigid building or other structure, that is, for structures with $n_1 \geq 1$ Hz. Flexible buildings or other structures are defined as those with $n_1 < 1$ Hz.

Once W_i is determined, the ice-covered structure is analyzed using the load combinations for strength design in ASCE/SEI 2.3 or the load combinations for allowable stress design in ASCE/SEI 2.4.

4.6   Design Temperatures for Freezing Rain

Some ice-sensitive structures can also be sensitive to changes in temperature. While maximum load effects usually occur at the lowest temperature when the structure is loaded with ice, it is possible for some types of structures, such as overhead cable systems, to experience maximum load effects at or around the melting point of ice, which is 32°F (0°C).

Temperatures concurrent with ice thickness due to freezing rain are given in ASCE/SEI Figures 10.6-1 and 10.6-2 for the contiguous 48 states and Alaska, respectively. The design temperature for ice and wind-on-ice is that from ASCE/SEI

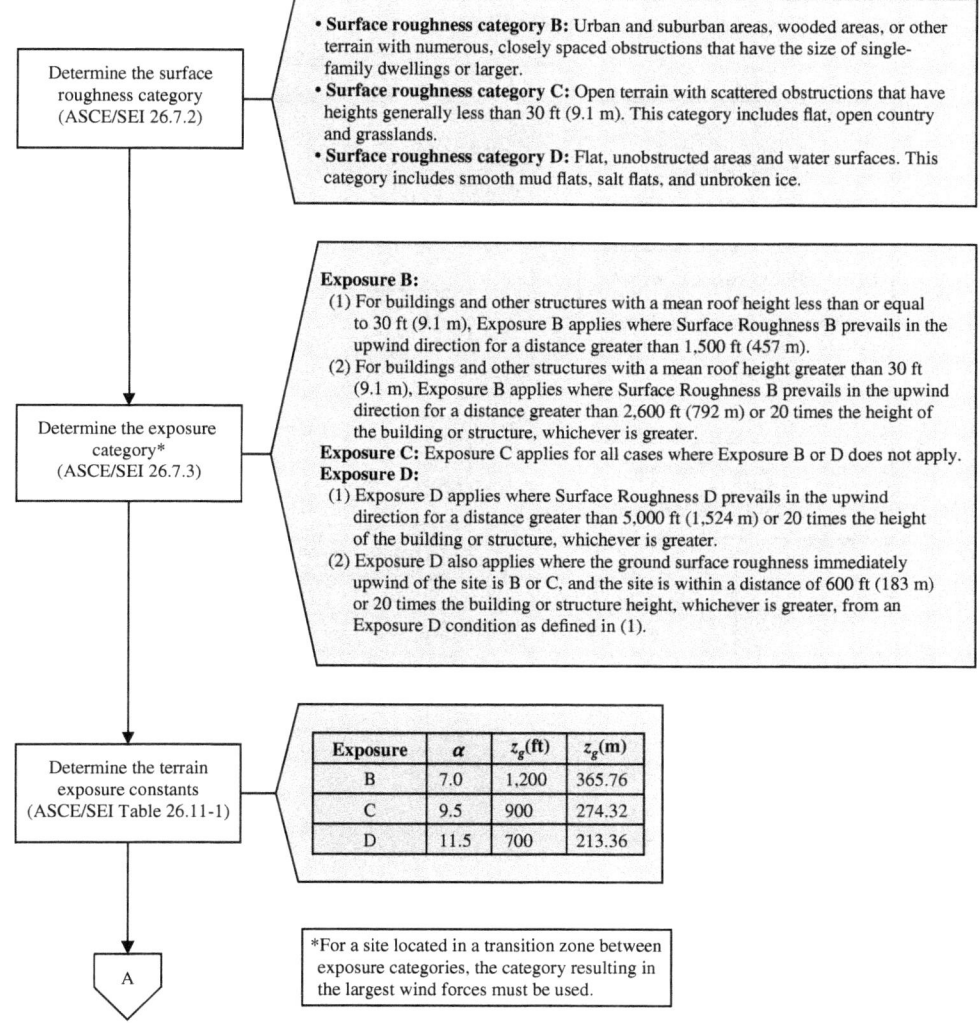

FIGURE 4.3 Flowchart to determine the wind velocity pressure, q_z.

Figures 10.6-1 and 10.6-2 or 32°F (0°C), whichever gives the maximum load effect. The design temperature for Hawaii is 32°F (0°C).

4.7 Partial Loading

Variations in ice thickness due to freezing rain at a given elevation are usually small over distances of about 1,000 ft (305 m). Thus, partial loading from freezing rain does not usually produce maximum load effects except in certain types of structures. Additional information on this topic can be found in ASCE/SEI C10.7.

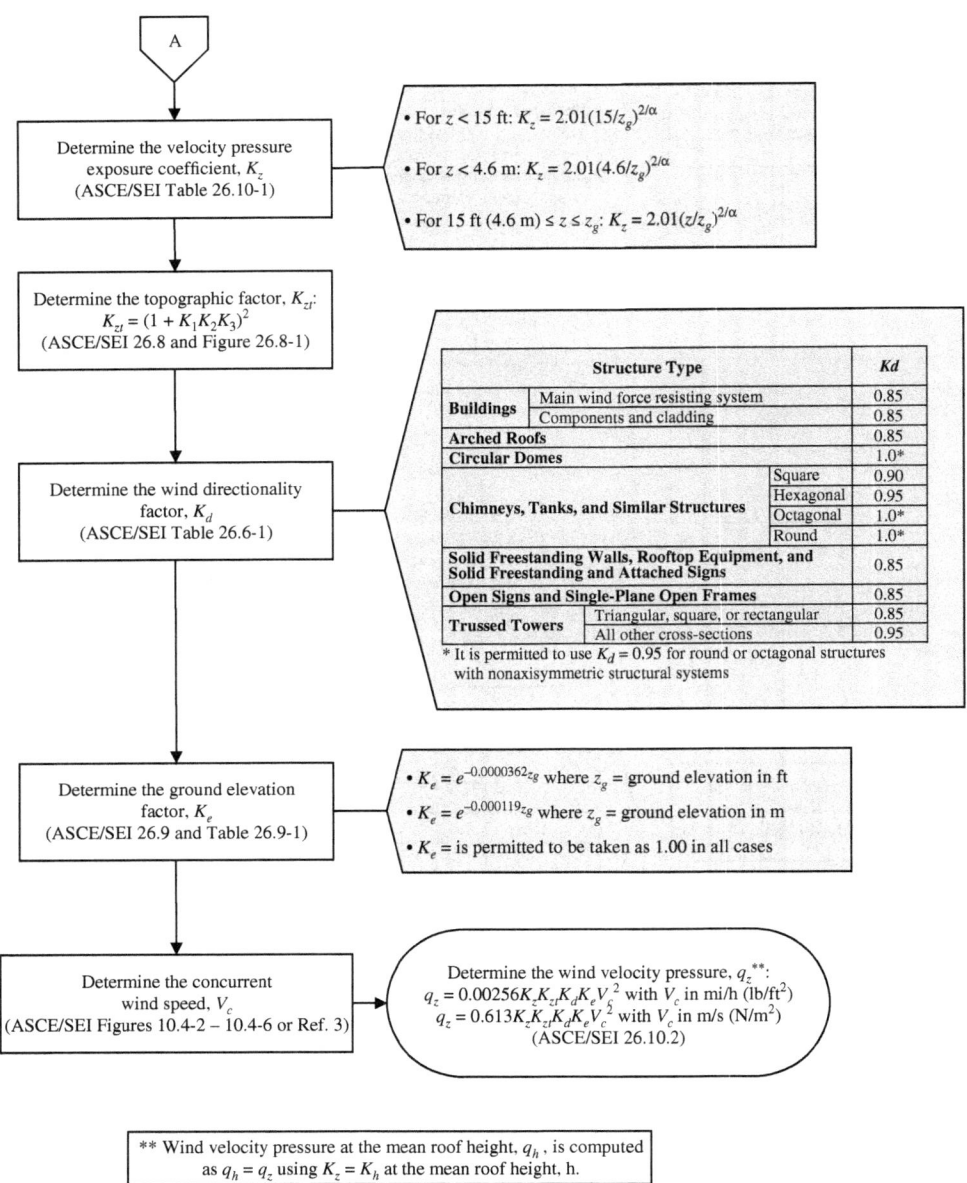

A

Determine the velocity pressure
exposure coefficient, K_z
(ASCE/SEI Table 26.10-1)

- For $z < 15$ ft: $K_z = 2.01(15/z_g)^{2/\alpha}$

- For $z < 4.6$ m: $K_z = 2.01(4.6/z_g)^{2/\alpha}$

- For 15 ft (4.6 m) $\leq z \leq z_g$: $K_z = 2.01(z/z_g)^{2/\alpha}$

Determine the topographic factor, K_{zt}:
$K_{zt} = (1 + K_1 K_2 K_3)^2$
(ASCE/SEI 26.8 and Figure 26.8-1)

Determine the wind directionality
factor, K_d
(ASCE/SEI Table 26.6-1)

Structure Type			Kd
Buildings	Main wind force resisting system		0.85
	Components and cladding		0.85
Arched Roofs			0.85
Circular Domes			1.0*
Chimneys, Tanks, and Similar Structures		Square	0.90
		Hexagonal	0.95
		Octagonal	1.0*
		Round	1.0*
Solid Freestanding Walls, Rooftop Equipment, and Solid Freestanding and Attached Signs			0.85
Open Signs and Single-Plane Open Frames			0.85
Trussed Towers	Triangular, square, or rectangular		0.85
	All other cross-sections		0.95

* It is permitted to use $K_d = 0.95$ for round or octagonal structures
with nonaxisymmetric structural systems

Determine the ground elevation
factor, K_e
(ASCE/SEI 26.9 and Table 26.9-1)

- $K_e = e^{-0.0000362 z_g}$ where z_g = ground elevation in ft

- $K_e = e^{-0.000119 z_g}$ where z_g = ground elevation in m

- K_e = is permitted to be taken as 1.00 in all cases

Determine the concurrent
wind speed, V_c
(ASCE/SEI Figures 10.4-2 – 10.4-6 or Ref. 3)

Determine the wind velocity pressure, q_z^{**}:
$q_z = 0.00256 K_z K_{zt} K_d K_e V_c^2$ with V_c in mi/h (lb/ft^2)
$q_z = 0.613 K_z K_{zt} K_d K_e V_c^2$ with V_c in m/s (N/m^2)
(ASCE/SEI 26.10.2)

** Wind velocity pressure at the mean roof height, q_h, is computed
as $q_h = q_z$ using $K_z = K_h$ at the mean roof height, h.

FIGURE 4.3   (Continued)

Structure	Force Coefficient, C_f	Wind-On-Ice Load, W_i (lb and N)	Notes
Chimneys, tanks, and similar structures (ASCE/SEI 10.5.1)	• For structures with square, hexagonal, and octagonal cross-sections, determine C_f from ASCE/SEI Figure 29.4-1. • For structures with round cross sections, determine C_f from ASCE/SEI Figure 29.4-1 for round cross-sections with $D\sqrt{q_z} \leq 2.5$ [in S.I.: $D\sqrt{q_z} \leq 5.3$] for all ice thicknesses, wind speeds, and structure diameters.	$q_z GC_f A_f$ [ASCE/SEI Equation (29.4-1)]	W_i is calculated based on the area of the structure, including ice, projected on a vertical plane normal to the wind direction. W_i is assumed to act parallel to the wind direction.
Solid freestanding walls and solid signs (ASCE/SEI 10.5.2)	Determine C_f from ASCE/SEI Figure 29.3-1 for Cases A, B, and C based on the dimensions of the wall or sign, including ice.	$q_h GC_f A_s$ [ASCE/SEI Equation (29.3-1)]	W_i is calculated for Cases A, B, and C in ASCE/SEI Figure 29.3-1. A_s is the gross area of the solid freestanding wall or freestanding solid sign, including ice.
Open signs and lattice frameworks* (ASCE/SEI 10.5.3)	• For flat members, determine C_f from ASCE/SEI Figure 29.4-2. • For rounded members and for the additional projected area caused by ice on both flat and rounded members, determine C_f from ASCE/SEI Figure 29.4-1 for rounded members with $D\sqrt{q_z} \leq 2.5$ [in S.I.: $D\sqrt{q_z} \leq 5.3$] for all ice thicknesses, wind speeds, and structure diameters.	$q_z GC_f A_f$ [ASCE/SEI Equation (29.4-1)]	W_i is calculated based on the area of all exposed members and elements, including ice, projected on a plane normal to the wind direction. W_i is assumed to act parallel to the wind direction. A_f is the solid area, including ice, projected normal to the wind direction.
Trussed towers* (ASCE/SEI 10.5.4)	Determine C_f from ASCE/SEI Figure 29.4-3. It is acceptable to reduce C_f for the additional projected area caused by ice on both round and flat members by the factor for rounded members in Note 3 of ASCE/SEI Figure 29.4-3.	$q_z GC_f A_f$ [ASCE/SEI Equation (29.4-1)]	W_i is to be applied in the directions resulting in maximum member forces and reactions. For all wind directions considered, A_f is the solid area of a tower face, including ice, projected on the plane of that face for the tower segment under consideration.
Guys and cables (ASCE/SEI 10.5.5)	Use $C_f = 1.2$.	$q_z GC_f A_f$ [ASCE/SEI Equation (29.4-1)]	See ASCE/SEI C10.5.5.

*The solidity ratio, ε, must be based on the projected area, including ice.

TABLE 4.3 Wind-On-Ice Loads, W_i

4.8 Examples

The following examples illustrate the determination of ice weight and wind-on-ice loads based on the provisions in ASCE/SEI Chapter 10. The overall design procedure in Fig. 4.1 is used as are the other resources in this chapter to determine the atmospheric ice loads.

4.8.1   Example 4.1—Calculation of Atmospheric Ice Loads for a Tank

Determine the ice weight and the wind-on-ice load for the circular reinforced concrete tank in Fig. 4.4. The tank is located at a relatively flat Exposure B site in Grand Rapids, MI. The tank does not contain hazardous, toxic, or explosive materials. Assume the natural fundamental frequency of the tank, n_1, is greater than 1 Hz.

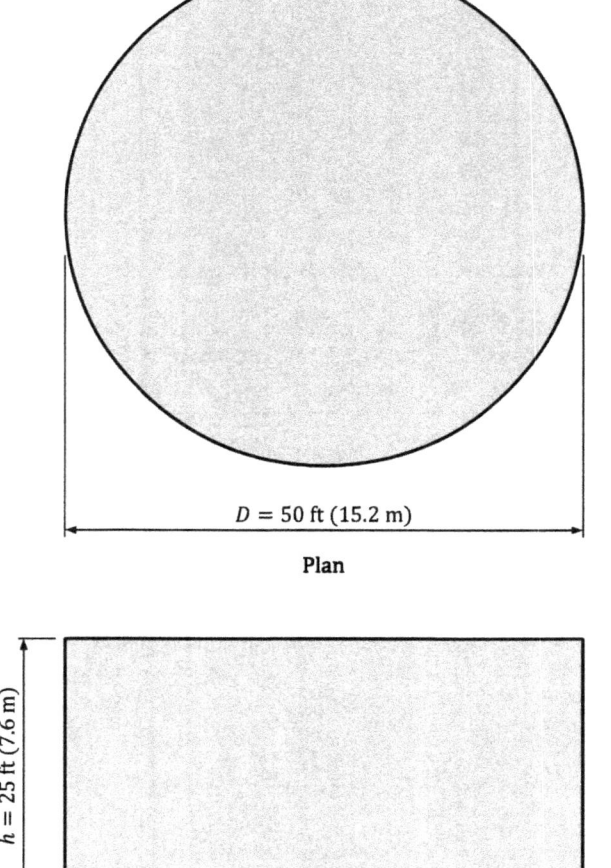

D = 50 ft (15.2 m)

Plan

h = 25 ft (7.6 m)

Elevation

FIGURE 4.4   Plan and elevation of the circular tank in Example 4.1.

Solution

Step 1—Determine the nominal ice thickness, t, and the concurrent wind speed, V_c Sec. 4.4.1

- The nominal ice thickness is equal to 1.5 in. (38 mm).   ASCE/SEI Figure 10.4-2
- The concurrent wind speed is equal to 40 mi/h (18 m/s).

Step 2—Determine the height factor, f_z   Sec. 4.4.2

$$\text{For } z = 25 \text{ ft: } f_z = \left(\frac{z}{33}\right)^{0.10} = \left(\frac{25}{33}\right)^{0.10} = 0.97 \tag{4.1}$$

$$\text{For } z = 7.6 \text{ m: } f_z = \left(\frac{z}{10}\right)^{0.10} = \left(\frac{7.6}{10}\right)^{0.10} = 0.97 \tag{4.2}$$

Step 3—Determine the importance factors I_i and I_w   Sec. 4.4.3

In accordance with ASCE/SEI Table 1.5-1, this tank can be classified under Risk Category II.

Therefore, $I_i = 1.0$ and $I_w = 1.0$.   ASCE/SEI Table 1.5-2

Step 4—Determine the topographic factor, K_{zt}   Sec. 4.4.4

Because the tank is located on a relatively flat site, $K_{zt} = 1.0$. ASCE/SEI 26.8.2

Step 5—Determine the design ice thickness, t_d   Sec. 4.4.5

$$t_d = tI_i f_z (K_{zt})^{0.35} = 1.5 \times 1.0 \times 0.97 \times (1.0)^{0.35} = 1.46 \text{ in.} \tag{4.4}$$

$$t_d = tI_i f_z (K_{zt})^{0.35} = 38 \times 1.0 \times 0.97 \times (1.0)^{0.35} = 37 \text{ mm}$$

Step 6—Determine the ice weight, D_i   Sec. 4.4.6

For a circular cross-section, $D_c = D = 50$ ft (15.2 m)   Fig. 4.2

$$A_i = \pi t_d (D_c + t_d) = \pi \times 1.46 \times [(50.0 \times 12) + 1.46] = 2{,}759 \text{ in.}^2$$

ASCE/SEI Equation (10.4-1)

$$D_i = 56A_i = 56 \times (2{,}759/144) = 1{,}073 \text{ lb/ft} \tag{Eq. (4.5)}$$

In. S.I.:

$$A_i = \pi t_d (D_c + t_d) = \pi \times 37 \times (15{,}200 + 37) = 1.771 \times 10^6 \text{ mm}^2$$

ASCE/SEI Equation (10.4-1)

$$D_i = 900gA_i = 900 \times 9.8 \times 1.771 = 15{,}620 \text{ N/m} \tag{4.6}$$

Alternatively, use the equations in Table 4.2:

$$D_i = 0.389\pi t_d (D_c + t_d) = 0.389 \times \pi \times 1.46 \times [(50.0 \times 12) + 1.46] = 1{,}073 \text{ lb/ft}$$

$$D_i = (8.82 \times 10^{-3})\pi t_d (D_c + t_d) = (8.82 \times 10^{-3}) \times \pi \times 37 \times (15{,}200 + 37) = 15{,}621 \text{ N/m}$$

Step 7—Determine the wind velocity pressure, q_z   Fig. 4.3

The design wind-on-ice load for tanks is determined by ASCE/SEI 29.4 (see Table 4.3). According to that section, q_z is evaluated at the centroid of the projected area normal to the wind, A_f. For this tank, the centroid occurs at the following height:

$$z = 25.0/2 = 12.5 \text{ ft (in S.I.: } z = 7.6/2 = 3.8 \text{ m).}$$

- Step 7a—Determine the exposure category ASCE/SEI 26.7.3

  From the design data, the exposure category is given as B.

- Step 7b—Determine the terrain exposure constants ASCE/SEI Table 26.11-1

  For Exposure B, $\alpha = 7.0$ and $z_g = 1,200$ ft (365.76 m).

- Step 7c—Determine the velocity pressure exposure coefficient, K_z

  ASCE/SEI Table 26.10-1

  For $z = 12.5$ ft (3.8 m): $K_z = 0.57$

- Step 7d—Determine the wind directionality factor, K_d ASCE/SEI Table 26.6-1

  For round tanks, $K_d = 1.0$ assuming the tank does not have a nonaxisymmetric structural system.

- Step 7e—Determine the ground elevation factor, K_e ASCE/SEI Table 26.9-1

  It is permitted to use $K_e = 1.0$ in all cases.

- Step 7f—Determine the wind velocity pressure, q_z ASCE/SEI 26.10.2

$$q_z = 0.00256 K_z K_{zt} K_d K_e V_c^2 \tag{4.9}$$

$$= 0.00256 \times 0.57 \times 1.0 \times 1.0 \times 1.0 \times 40^2 = 2.3 \text{ lb/ft}^2$$

$$q_z = 0.613 K_z K_{zt} K_d K_e V_c^2 \tag{4.10}$$

$$= 0.613 \times 0.57 \times 1.0 \times 1.0 \times 1.0 \times 18^2 = 113.2 \text{ N/m}^2$$

Step 8—Determine the gust-effect factor, G Sec. 4.5.2

It is given in the design data that $n_1 > 1$ Hz, which means the tank can be classified as rigid.

For rigid structures, G is permitted to be taken as 0.85 (ASCE/SEI 26.11.1).

Step 9—Determine the wind force coefficient, C_f Sec. 4.5.2

The wind force coefficient C_f for a round tank is determined in accordance with ASCE/SEI Figure 29.4-1 corresponding to $D\sqrt{q_z} \leq 2.5$ (in S.I.: $D\sqrt{q_z} \leq 0.53$) for all ice thicknesses, wind speeds, and structure diameters (ASCE/SEI 10.5.1; see Table 4.3).

For $h/D = 0.5$, use $C_f = 0.7$.

Step 10—Determine the wind-on-ice load, W_i Sec. 4.5.2

A_f = projected area normal to the wind including ice

$$= [50.0 + (2 \times 1.46/12)] \times [25.0 + (1.46/12)] = 1,262 \text{ ft}^2$$

$$W_i = F = q_z GC_f A_f = 2.3 \times 0.85 \times 0.7 \times 1,262 = 1,727 \text{ lb} \qquad \text{Table 4.3}$$

In S.I.:

A_f = projected area normal to the wind including ice

$$= [15.2 + (2 \times 37/1,000)] \times [7.6 + (37/1,000)] = 116.6 \text{ m}^2$$

$$W_i = F = q_z GC_f A_f = 113.2 \times 0.85 \times 0.7 \times 116.6 = 7,853 \text{ N} \qquad \text{Table 4.3}$$

This force is applied at the centroid of the projected area normal to the wind, A_f, which is 12.5 ft (3.8 m) above ground level and 25 ft (7.6 m) from either side of the tank.

4.8.2 Example 4.2—Calculation of Atmospheric Ice Loads for a Chimney

Determine the ice weight and the wind-on-ice load for a round reinforced concrete chimney 70 ft (21.3 m) tall with an outside diameter of 7 ft (2.1 m). The chimney is located at a relatively flat Exposure B site in Minneapolis, MN. Assume the natural fundamental frequency of the chimney, n_1, is greater than 1 Hz.

Solution

Step 1—Determine the nominal ice thickness, t, and the concurrent wind speed, V_c Sec. 4.4.1

The nominal ice thickness is equal to 1.5 in. (38 mm). ASCE/SEI Figure 10.4-2

The concurrent wind speed is equal to 50 mi/h (22 m/s).

Step 2—Determine the height factor, f_z Sec. 4.4.2

$$\text{For } z = 70 \text{ ft: } f_z = \left(\frac{z}{33}\right)^{0.10} = \left(\frac{70}{33}\right)^{0.10} = 1.08 \tag{4.1}$$

$$\text{For } z = 21.3 \text{ m: } f_z = \left(\frac{z}{10}\right)^{0.10} = \left(\frac{21.3}{10}\right)^{0.10} = 1.08 \tag{4.2}$$

Step 3—Determine the importance factors I_i and I_w Sec. 4.4.3

In accordance with ASCE/SEI Table 1.5-1, this chimney can be classified under Risk Category II.

Therefore, $I_i = 1.0$ and $I_w = 1.0$. ASCE/SEI Table 1.5-2

Step 4—Determine the topographic factor, K_{zt} Sec. 4.4.4

Because the chimney is located on a relatively flat site, $K_{zt} = 1.0$.

ASCE/SEI 26.8.2

Step 5—Determine the design ice thickness, t_d Sec. 4.4.5

$$t_d = t I_i f_z (K_{zt})^{0.35} = 1.5 \times 1.0 \times 1.08 \times (1.0)^{0.35} = 1.62 \text{ in.} \tag{4.4}$$

$$t_d = t I_i f_z (K_{zt})^{0.35} = 38 \times 1.0 \times 1.08 \times (1.0)^{0.35} = 41 \text{ mm}$$

Step 6—Determine the ice weight, D_i Sec. 4.4.6

For a circular cross-section, $D_c = D = 7$ ft (2.1 m) Fig. 4.2

$$A_i = \pi t_d (D_c + t_d) = \pi \times 1.62 \times [(7.0 \times 12) + 1.62] = 436 \text{ in.}^2$$

ASCE/SEI Equation (10.4-1)

$$D_i = 56 A_i = 56 \times (436/144) = 170 \text{ lb/ft} \tag{Eq. (4.5)}$$

In S.I.:

$$A_i = \pi t_d (D_c + t_d) = \pi \times 41 \times (2,100 + 41) = 0.276 \times 10^6 \text{ mm}^2$$

ASCE/SEI Equation (10.4-1)

$$D_i = 900 g A_i = 900 \times 9.8 \times 0.276 = 2,434 \text{ N/m} \tag{4.6}$$

Alternatively, use the equations in Table 4.2:

$$D_i = 0.389\pi t_d (D_c + t_d) = 0.389 \times \pi \times 1.62 \times [(7.0 \times 12) + 1.62] = 170 \text{ lb/ft}$$

$$D_i = (8.82 \times 10^{-3})\pi t_d (D_c + t_d) = (8.82 \times 10^{-3}) \times \pi \times 41 \times (2,100 + 41) = 2,432 \text{ N/m}$$

Step 7—Determine the wind velocity pressure, q_z   Fig. 4.3

The design wind-on-ice load for chimneys is determined by ASCE/SEI 29.4 (see Table 4.3). According to that section, q_z is evaluated at the centroid of the projected area normal to the wind, A_f. For this chimney, the centroid occurs at the following height:

$$z = 70/2 = 35 \text{ ft [in S.I.: } z = 21.3/2 = 10.7 \text{ m]} \cdot$$

- Step 7a—Determine the exposure category   ASCE/SEI 26.7.3

  From the design data, the exposure category is given as B.

- Step 7b—Determine the terrain exposure constants   ASCE/SEI Table 26.11-1

  For Exposure B, $\alpha = 7.0$ and $z_g = 1,200$ ft (365.76 m).

- Step 7c—Determine the velocity pressure exposure coefficient, K_z

  ASCE/SEI Table 26.10-1

  $$\text{For } z = 35 \text{ ft: } K_z = 2.01(z/z_g)^{2/\alpha} = 2.01 \times (35/1,200)^{2/7.0} = 0.73$$
  $$\text{For } z = 10.7 \text{ m: } K_z = 2.01(z/z_g)^{2/\alpha} = 2.01 \times (10.7/365.76)^{2/7.0} = 0.73$$

- Step 7d—Determine the wind directionality factor, K_d ASCE/SEI Table 26.6-1

  For round chimneys, $K_d = 1.0$ assuming the chimney does not have a nonaxisymmetric structural system.

- Step 7e—Determine the ground elevation factor, K_e ASCE/SEI Table 26.9-1

  It is permitted to use $K_e = 1.0$ in all cases.

- Step 7f—Determine the wind velocity pressure, q_z ASCE/SEI 26.10.2

  $$q_z = 0.00256 K_z K_{zt} K_d K_e V_c^2 \qquad (4.9)$$
  $$= 0.00256 \times 0.73 \times 1.0 \times 1.0 \times 1.0 \times 50^2 = 4.7 \text{ lb/ft}^2$$

  $$q_z = 0.613 K_z K_{zt} K_d K_e V_c^2 \qquad (4.10)$$
  $$= 0.613 \times 0.73 \times 1.0 \times 1.0 \times 1.0 \times 22^2 = 216.6 \text{ N/m}^2$$

Step 8—Determine the gust-effect factor, G   Sec. 4.5.2

It is given in the design data that $n_1 > 1$ Hz, which means the chimney can be classified as rigid.

For rigid structures, G is permitted to be taken as 0.85 (ASCE/SEI 26.11.1).

Step 9—Determine the wind force coefficient, C_f   Sec. 4.5.2

The wind force coefficient C_f for a round chimney is determined in accordance with ASCE/SEI Figure 29.4-1 corresponding to $D\sqrt{q_z} \leq 2.5$ (in S.I.: $D\sqrt{q_z} \leq 0.53$)

for all ice thicknesses, wind speeds, and structure diameters (ASCE/SEI 10.5.1; see Table 4.3).

For $h/D = 10$, use $C_f = 0.9$ by linear interpolation.

Step 10—Determine the wind-on-ice load, W_i   Sec. 4.5.2

A_f = projected area normal to the wind including ice

$$= [7.0 + (2 \times 1.62/12)] \times [70.0 + (1.62/12)] = 510 \text{ ft}^2$$

$W_i = F = q_z G C_f A_f = 4.7 \times 0.85 \times 0.9 \times 510 = 1,834 \text{ lb}$ Table 4.3

In S.I.:

A_f = projected area normal to the wind including ice

$$= [2.1 + (2 \times 41/1,000)] \times [21.3 + (41/1,000)] = 46.6 \text{ m}^2$$

$W_i = F = q_z G C_f A_f = 216.6 \times 0.85 \times 0.9 \times 46.6 = 7,722 \text{ N}$ Table 4.3

This force is applied at the centroid of the projected area normal to the wind, A_f, which is 35.0 ft (10.7 m) above ground level and 3.5 ft (1.1 m) from either side of the chimney.

4.8.3   Example 4.3—Calculation of Atmospheric Ice Loads for a Solid Freestanding Wall

Determine the ice weight and the wind-on-ice load for an architectural freestanding screen wall 15 ft (4.6 m) tall and 50 ft (15.2 m) long. The screen wall is 10 percent open and is located at a relatively flat Exposure C site in Boston, MA. Assume the natural fundamental frequency of the wall, n_1, is greater than 1 Hz.

Solution

Step 1—Determine the nominal ice thickness, t, and the concurrent wind speed, V_c

Sec. 4.4.1

The nominal ice thickness is equal to 1.0 in. (25 mm).   ASCE/SEI Figure 10.4-2

The concurrent wind speed is equal to 50 mi/h (22 m/s).

Step 2—Determine the height factor, f_z   Sec. 4.4.2

$$\text{For } z = 15 \text{ ft: } f_z = \left(\frac{z}{33}\right)^{0.10} = \left(\frac{15}{33}\right)^{0.10} = 0.92 \tag{4.1}$$

$$\text{For } z = 4.6 \text{ m: } f_z = \left(\frac{z}{10}\right)^{0.10} = \left(\frac{4.6}{10}\right)^{0.10} = 0.92 \tag{4.2}$$

Step 3—Determine the importance factors I_i and I_w   Sec. 4.4.3

This wall represents a low risk to human life in the event of failure. In accordance with ASCE/SEI Table 1.5-1, this chimney can be classified under Risk Category I.

Therefore, $I_i = 0.8$ and $I_w = 1.0$.   ASCE/SEI Table 1.5-2

Step 4—Determine the topographic factor, K_{zt}   Sec. 4.4.4

Because the wall is located on a relatively flat site, $K_{zt} = 1.0$. ASCE/SEI 26.8.2

Step 5—Determine the design ice thickness, t_d   Sec. 4.4.5

$$t_d = t I_i f_z (K_{zt})^{0.35} = 1.0 \times 0.8 \times 0.92 \times (1.0)^{0.35} = 0.74 \text{ in.} \qquad (4.4)$$

$$t_d = t I_i f_z (K_{zt})^{0.35} = 25 \times 0.8 \times 0.92 \times (1.0)^{0.35} = 18 \text{ mm}$$

Step 6—Determine the ice weight, D_i   Sec. 4.4.6

For a flat plate, A_s is the area of one side of the plate: ASCE/SEI 10.4.1

$$A_s = 50 \times 15 = 750 \text{ ft}^2$$

The volume of ice, V_i, is equal to the following for a vertical plate:

$$V_i = 0.8 \pi t_d A_s = 0.8 \times \pi \times (0.74/12) \times 750 = 116.2 \text{ ft}^3$$
$$D_i = 56 V_i = 56 \times 116.2 = 6,507 \text{ lb}$$

In S.I.:

For a flat plate, A_s is the area of one side of the plate: ASCE/SEI 10.4.1

$$A_s = 15.2 \times 4.6 = 69.9 \text{ m}^2$$

The volume of ice, V_i, is equal to the following for a vertical plate:

$$V_i = 0.8 \pi t_d A_s = 0.8 \times \pi \times (18/1,000) \times 69.9 = 3.2 \text{ m}^3$$
$$D_i = 900 g V_i = 900 \times 9.8 \times 3.2 = 28,224 \text{ N}$$

Alternatively, use the equations in Table 4.2:

$$D_i = 3.73 \pi t_d bh = 3.73 \times \pi \times 0.74 \times 50 \times 15 = 6,504 \text{ lb}$$
$$D_i = 7.06 \pi t_d bh = 7.06 \times \pi \times 18 \times 15.2 \times 4.6 = 27,914 \text{ N}$$

Step 7—Determine the wind velocity pressure, q_z   Fig. 4.3

The design wind-on-ice load for solid freestanding walls is determined by ASCE/SEI 29.3 (see Table 4.3). According to that section, $q_z = q_h$ is evaluated at the top of the wall where $z = 15$ ft (4.6 m).

- Step 7a—Determine the exposure category   ASCE/SEI 26.7.3

  From the design data, the Exposure Category is given as C.

- Step 7b—Determine the terrain exposure constants   ASCE/SEI Table 26.11-1

  For Exposure C, $\alpha = 9.5$ and $z_g = 900$ ft (274.32 m).

- Step 7c—Determine the velocity pressure exposure coefficient, K_z

  ASCE/SEI Table 26.10-1

  For $z = 15$ ft (4.6 m): $K_z = 0.85$

- Step 7d—Determine the wind directionality factor, K_d ASCE/SEI Table 26.6-1

  For solid freestanding walls, $K_d = 0.85$.

- Step 7e—Determine the ground elevation factor, K_e ASCE/SEI Table 26.9-1

  It is permitted to use $K_e = 1.0$ in all cases.

- Step 7f—Determine the wind velocity pressure, q_z ASCE/SEI 26.10.2

$$q_z = q_h = 0.00256 K_z K_{zt} K_d K_e V_c^2 \tag{4.9}$$
$$= 0.00256 \times 0.85 \times 1.0 \times 0.85 \times 1.0 \times 50^2 = 4.6 \ \text{lb/ft}^2$$

$$q_z = q_h = 0.613 K_z K_{zt} K_d K_e V_c^2 \tag{4.10}$$
$$= 0.613 \times 0.85 \times 1.0 \times 0.85 \times 1.0 \times 22^2 = 214.4 \ \text{N/m}^2$$

Step 8—Determine the gust-effect factor, G   Sec. 4.5.2

It is given in the design data that $n_1 > 1$ Hz, which means the wall can be classified as rigid.

For rigid structures, G is permitted to be taken as 0.85 (ASCE/SEI 26.11.1).

Step 9—Determine the wind force coefficient, C_f   Sec. 4.5.2

The wind force coefficient C_f for a solid freestanding wall is determined in accordance with ASCE/SEI Figure 29.3-1. In general, cases A, B, and C must be considered.

In this example, the aspect ratio $B/s = 50/15 = 3.3 > 2$ (in S.I.: $15.2/4.6 = 3.3 > 2$). Therefore, in accordance with Note 3 in ASCE/SEI Figure 29.3-1, all three load cases must be considered.

Cases A and B
For $s/h = 1$ and $B/S = 3.3$, $C_f = 1.37$ by linear interpolation.

According to Note 2 in ASCE/SEI Figure 29.3-1, force coefficients for solid freestanding walls with openings may be multiplied by the reduction factor $[1 - (1 - \varepsilon)^{1.5}]$ where ε is equal to the ratio of the solid area to the gross area of the wall, including ice. In lieu of determining ε including ice, do not reduce C_f by this reduction factor.

Case C
The following force coefficients are obtained from ASCE/SEI Figure 29.3-1 by linear interpolation [which includes the $(1.8 - s/h) = 1.8 - 1.0 = 0.8$ reduction factor in Note 4 of ASCE/SEI Figure 29.3-1]:

From 0 to 15 ft (4.6 m): $C_f = 0.8 \times 2.69 = 2.15$

From 15 ft (4.6 m) to 30 ft (9.1 m): $C_f = 0.8 \times 1.76 = 1.41$

From 30 ft (9.1 m) to 50 ft (15.2 m): $C_f = 0.8 \times 1.20 = 0.96$

Step 10—Determine the wind-on-ice load, W_i   Sec. 4.5.2

For cases A and B:

A_s = gross area of the wall including ice
$$= [50.0 + (2 \times 0.74/12)] \times [15.0 + (0.74/12)] = 755 \ \text{ft}^2$$
$$W_i = F = q_h G C_f A_s = 4.6 \times 0.85 \times 1.37 \times 755 = 4,044 \ \text{lb} \qquad \text{Table 4.3}$$

In case A with $s/h = 1$, this force is applied at a distance of $0.05h = 0.05 \times 15 = 0.75$ ft above the geometric center of the wall, that is, $(15/2) + 0.75 = 8.25$ ft above the ground level (see ASCE/SEI Figure 29.3-1).

In case B, this force is applied 8.25 ft above ground level and $(50/2) - (0.2 \times 50) = 15$ ft from the windward edge of the wall (see ASCE/SEI Figure 29.3-1).

For case C:

In the region between 0 and 15 ft from the windward edge of the wall:

$$A_s = [15 + (0.74/12)] \times [15 + (0.74/12)] = 227 \text{ ft}^2$$

$$W_i = F = q_h GC_f A_s = 4.6 \times 0.85 \times 2.15 \times 227 = 1,908 \text{ lb}$$

This force is applied 8.25 ft above ground level and $15/2 = 7.5$ ft from the windward edge of the wall.

In the region between 15 ft and 30 ft from the windward edge of the wall:

$$A_s = 15 \times [15 + (0.74/12)] = 226 \text{ ft}^2$$

$$W_i = F = q_h GC_f A_s = 4.6 \times 0.85 \times 1.41 \times 226 = 1,246 \text{ lb}$$

This force is applied 8.25 ft above ground level and $15 + (15/2) = 22.5$ ft from the windward edge of the wall.

In the region between 30 ft and 50 ft from the windward edge of the wall:

$$A_s = [20 + (0.74/12)] \times [15 + (0.74/12)] = 302 \text{ ft}^2$$

$$W_i = F = q_h GC_f A_s = 4.6 \times 0.85 \times 0.96 \times 302 = 1,134 \text{ lb}$$

This force is applied 8.25 ft above ground level and $30 + (20/2) = 40$ ft from the windward edge of the wall.

In S.I.:

For cases A and B:

A_s = gross area of the wall including ice

$$= [15.2 + (2 \times 18/1,000)] \times [4.6 + (18/1,000)] = 70.4 \text{ m}^2$$

$$W_i = F = q_h GC_f A_s = 214.4 \times 0.85 \times 1.37 \times 70.4 = 17,577 \text{ N} \qquad \text{Table 4.3}$$

In case A with $s/h = 1$, this force is applied at a distance of $0.05h = 0.05 \times 4.6 = 0.23$ m above the geometric center of the wall, that is, $(4.6/2) + 0.23 = 2.53$ m above the ground level (see ASCE/SEI Figure 29.3-1).

In case B, this force is applied 2.53 m above ground level and $(15.2/2) - (0.2 \times 15.2) = 4.6$ m from the windward edge of the wall (see ASCE/SEI Figure 29.3-1).

For case C:

In the region between 0 and 4.6 m from the windward edge of the wall:

$$A_s = [4.6 + (18 / 1,000)] \times [4.6 + (18/1,000)] = 21.3 \text{ m}^2$$

$$W_i = F = q_h GC_f A_s = 214.4 \times 0.85 \times 2.15 \times 21.3 = 8,346 \text{ N}$$

This force is applied 2.53 m above ground level and $4.6/2 = 2.3$ m from the windward edge of the wall.

In the region between 4.6 m and 9.1 m from the windward edge of the wall:

$$A_s = 4.6 \times [4.6 + (18/1,000)] = 21.2 \text{ m}^2$$

$$W_i = F = q_h GC_f A_s = 214.4 \times 0.85 \times 1.41 \times 21.2 = 5,448 \text{ N}$$

This force is applied 2.53 m above ground level and $4.6 + (4.6/2) = 6.9$ m from the windward edge of the wall.

In the region between 9.1 m and 15.2 m from the windward edge of the wall:

$$A_s = [6.1 + (18/1,000)] \times [4.6 + (18/1,000)] = 28.3 \text{ m}^2$$

$$W_i = F = q_h GC_f A_s = 214.4 \times 0.85 \times 0.96 \times 28.3 = 4,951 \text{ N}$$

This force is applied 2.53 m above ground level and $9.2 + (6.1/2) = 12.3$ m from the windward edge of the wall.

CHAPTER 5

References

1. International Code Council. 2017. *2018 International Building Code,* Washington, DC.
2. Structural Engineering Institute of the American Society of Civil Engineers (ASCE). 2017. *Minimum Design Loads and Associated Criteria for Buildings and Other Structures*, ASCE/SEI 7-16, Reston, VA.
3. American Society of Civil Engineers (ASCE). 2018. ASCE 7 Hazard Tool. https://asce7hazardtool.online/.
4. National Oceanic and Atmospheric Administration (NOAA). 2017. Precipitation Frequency Data Server (PFDS). https://hdsc.nws.noaa.gov/.
5. Factory Mutual Insurance Company. 2016. *Roof Loads for New Construction*, FM Global Property Loss Prevention Data Sheet 1-54. Johnston, RI.
6. van Herwijnen, F., Snijder, H.H., and Fijneman, H.J. 2006. "Structural Design for Ponding of Rainwater on Roof Structures." *HERON*, 51(2/3): 115-150.
7. Applied Technology Council (ATC). ATC Hazards by Location. 2020. http://hazards.atcouncil.org.